THE CFZ YEARBOOK 2000/1

edited by

JONATHAN DOWNES
RICHARD FREEMAN

Typeset by Jonathan Downes,
Cover and Layout by Hennis/Spider inc for CFZ Communications
Using Microsoft Word 2000, Microsoft , Publisher 2000, Adobe Photoshop CS.

Photographs © 2008 CFZ except where noted

First published in Great Britain by CFZ Press

CFZ Press
Myrtle Cottage
Woolsery
Bideford
North Devon
EX39 5QR

THE CENTRE FOR FORTEAN ZOOLOGY
www.cfz.org.uk

© CFZ MMVIII

ISBN: 978-1-905723-28-7

Contents

Introduction

To the 2008 Edition

Dear friends,

I have enjoyed the re-mastering process on this volume immensely, and - I have to admit - considerably more than I did some of the other ones in the series. Despite our brave plan at the beginning to publish a yearbook every year, in 2000 we singularly failed to do so. It was a turbulent time for us at the CFZ, and especially for me personally, and it was only really when re-reading the catalogue of woes that had befallen us, when proofreading the article in this volume which describes events at the CFZ as we spanned the millennium, that I realised what a horrible time we actually had been through.

However, by early 2001 we had picked ourselves up, dusted ourselves down, and - to a certain extent at least - were getting on with our lives once again.

There are two main differences between the CFZ before the millennium, and the CFZ from 2001 onwards. The first of these is that the CFZ of the 21st Century were more than averagely computer literate, whereas prior to 2000 we really had no idea of what we were doing. This had a number of knock-on effects for us, most notably that we were now subscribed to a large number of news-feeds, newsgroups, and discussion forums, so we were receiving, and therefore disseminating the vast majority of cryptozoological news from around the world, whereas in the early years we were dependent almost entirely on our network of regional representatives.

The second main difference was that following our inaugural *Weird Weekend* in May 2000, we had become more and more involved, both socially and professionally, with researchers from around the world, and from throughout the gamut of the fortean universe. People like Nick Redfern, Tim Matthews, Matthew Williams, and Mike Hallowell became personal friends and we worked together with them on various projects. The CFZ was no longer a bunch of slightly strange people in Exeter working in splendid isolation.

But on the down side, although we had big plans, and there were many things that we wanted to do, we were living in abject poverty and had very little money left with which to do them.

When I sat down to write my autobiography, *Monster Hunter* in 2004, I had a very embarrassing experience. When I tried to remember what I had done in the year 2001, I could remember very little. Great swathes of time were blank.

I was accused at the time of having followed in the not inconsiderable footsteps of Joe Walsh, who claimed much the same thing about large lumps of his career, and I have to admit that this is indeed partially, because I spent the first few months of that year finally overcoming my drug problem, but it is also that most of what we did involved sitting down and writing.

Richard and I were both involved in major writing projects which were not to come to fruition for several more years. I was writing the early chapters of what would become *Monster Hunter*, and Richard was engaged on *Dragons: More than a Myth?* Both books finally weighed in at over 270,000 words and took years to write.

In addition, our only funding came from writing for various magazines, most of which weren't in the slightest bit fortean, and this book was about the only major cryptozoological achievement of the year. We managed a few minor investigations, mostly big cat related, and all local, and we promoted our second *Weird Weekend*, but most of the time we just wrote.

Sitting down, and remastering this volume has been a very pleasant experience, because it is good to discover quite what a high standard we were starting to aspire to with the CFZ Yearbooks. This was the first year that the production of the Yearbook became one of the major events of the annual CFZ calendar, (as it has done ever since). As there hadn't been a 2000 yearbook, this volume covered both years and contained some fine articles. Dr Karl Shuker did us proud with his article about tinamous in the United Kingdom, Richard's piece of historical detective work about `Trader` Horn was particularly enjoyable, and, re-reading it for the first time since I wrote it, I believe that my paper on the changing mammalogy of Hong Kong has certain things to recommend it.

Something else that can be seen from this volume is that all of the authors who contributed to this volume are still working with us, but that they all have (or will have by the end of 2008) had books published by CFZ Press.

This is a direct result of our activities back in 2001, because this was the year that we finally gave up trying to court the mainstream media, and decided to "go it alone".

For years, Richard and I had been trying to get agents, and publish books within the mainstream of the publishing industry. And for years we had been politely, and not so politely, rebuffed. The mainstream publishing industry were not interested in us, and in many cases seemed to dislike us intensely. This was especially galling, because at the time all sorts of ridiculous books about UFOs, and other aspects of the fortean omniverse were being published, often by people who quite simply had only the most passing acquaintance with the Queen's English, and so were not just talking nonsense, but barely-literate nonsense which is considerably worse.

On top of this, these people were getting substantial advances for their work. I had managed to sell four books; two each to two different publishers. The largest advance I got was a couple of grand. The *en dit* around the fortean hack writers of the time was that one writer (who shall remain nameless) got a £200,000 advance for one particular book about aliens in 1998.

By 2001 we realised that we would never enter this not very exclusive club, and made plans to go it alone with our own publishing company. This was the first result of our decision, and I think that on the whole we did pretty well.

Enjoy.

Jon Downes
North Devon
May 3 2008

Introduction

BY

Jonathan Downes

Dear Friends,

Welcome to another edition of the CFZ Yearbook. As always, this is intended as a forum for longer articles and research papers that would not fit comfortably within the format of *Animals & Men,* the journal of the Centre for Fortean Zoology.

This is our fifth yearbook, and after a gap last year it is good to be back. Essentially, we decided that with everybody and his aunt issuing special "Millenium" books we would, as per usual, buck the trend and wait a year before we put out another volume.

I am pleased to say that, in our opinion at least, it has been very much worth the wait. In fact, we had so much material available that we are already thinking in terms of next year`s volume!

The theme for this year seems to be dragons. This was not an intentional move on our behalf, but merely the way that it worked out. However, with such a noted Draconologist as Richard Freeman, having been on the full time faculty of the CFZ for something over three years now, it was, perhaps, inevitable, that eventually one of the Yearbooks would be Draconian in viewpoint.

As well as the stalwarts of the CFZ Yearbooks over the years, like Dr Karl Shuker, and Neil Arnold, we are proud to be able to introduce you to someone making his Yearbook debut. Mike Hallowell is a respected investigator and author from Tyneside. I met him first during my unfortunate sojourn as editor of *Quest* Magazine, and we became firm friends. I am sure that his paper on the dragons of the North East of England will be only the first of many appearances he will make in these pages.

Never afraid to shrink from controversy we are also pleased to present an article by our old friend Chris Moiser on the history of humans being kept in captivity and exhibited in British zoos and circuses. This is an unfortunate, and often unjustly ignored footnote to British zoological history. However it should be in the public domain and I am proud that the CFZ is prepared to publish material like this, which would, I am certain, be too sensitive for many other organisations or publications to tackle.

I hope we shall continue to deal with topics other organisations are not prepared to approach.

Enjoy,

Jon Downes
(Director, The Centre for Fortean Zoology)
Exeter.
April 2001

Kent's Boar War

by

Neil Arnold

A dangerous breed of animal, not seen in the south-east countryside for 300 years, is back and thriving.

In *Animals & Men* a year or so ago I reported on wild boar being once again introduced into the regions woodland, as well as some clusters existing in the wild due to the great storm of '87.

Back then the existence of such a creature was not an overwhelming problem, but now it is speculated that a population of over one-thousand is living in the Weald and causing havoc.

Not only are crops and fields decimated, but also the boar are now considered a threat to humans. Indeed, the problem has become so severe that in June of 1999 various landowners decided it was time to organise an emergency mass meeting.

Was this action a natural and justifiable reaction? Or merely man's usual way of destroying nature and its creations? I am very much 'for' the countryside, even if it means us humans becoming accustomed to big cats and other out of place animals roaming rural areas. I also believe that when a once-native creature gradually makes its way back home, it should be welcomed with open arms, even if its presence alarms the dog-walkers and picnic lovers amongst us.

The wild boar is not a pleasant creature. The handfuls that escaped various farms during 1987 struck fear into ramblers and farmers. Folk had reason to be frightened, especially as one man saw his three large dogs ripped to shreds by a marauding specimen only a couple of years ago. However, the rare occasion of friction now seems to have been blown out of proportion as various campaigners now suggest all-out war in order to curb the 'nuisance' that is spreading rapidly.

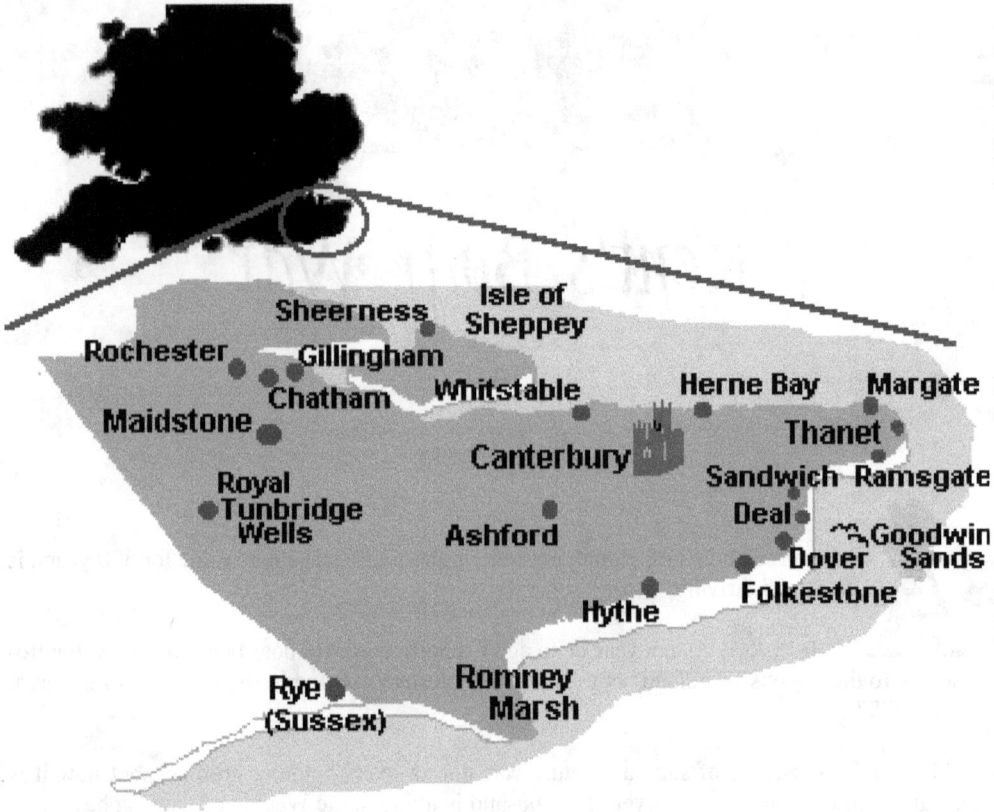

He told me: *"Farmers have been attacked in combine harvesters and others have been chased from their property, and the day will come when someone is killed."* The fact that no-one has ever been attacked seems to point to an over-reaction by a group who are to lobby the government to tackle the problem.

The wild boar summit took place at Maidstone's Tudor Park Hotel and was organised by accountant Paul Edgson Wright, who has a keen interest in agricultural matters. A government study is underway but it has yet officially accepted that the boar has come to stay. However, video footage shown at the meeting clearly shows boar settled into Kent and the Sussex border. One such film showed a twelve-month-old boar, weighing around 120 pounds. It is allegedly one of sixty living in the Beckley area, where there is speculation that there are males of 450 pounds in weight, and sows around 350 lbs.

In fact, a number of daytime films depicted the tusked terrors plodding through open farmland, but it is their nocturnal habits that put farmers out of business and people in danger. Fair enough, one swipe of a mighty tooth could sever a human, but I feel that there must be another way to restrict the growth of the boar.

Hunter and conservationist Angus Irvine said: *"They don't exist in law and the government has to decide whether they exist. Maybe then we can make decisions from that point."*

However, rational thinking seems too out of the spectrum for some campaigners. It seems that whilst the opposition is condemning the wild boar, they ought to look at themselves and maybe put their own spears away instead of acting like primitive hunters looking to destroy anything that appears as a threat.

Such radical opinions provoked a poster campaign that was obviously intended to shock people and to turn them against a creature that has only done harm to humans in various European countries and not in Kent.

Even then, when sifting through the facts, I notice that in each attack, the human had intruded upon a boar and its young. So, is the creature really so savage? Not really, especially considering that the woodlands of Britain were once its own, that is until we shunted them out and played egotistical hunting games, before serving them up on a nice hot plate.

The mock poster read:

WILD BOAR KILLS MOTHER AND DAUGHTER:

A mother and daughter were killed while walking through an English woodland along a public footpath. Their dog, it is believed, disturbed the boar, which charged out of the undergrowth killing the young girl instantly, and so severely wounding the mother by ripping with its tusks, that she subsequently died.

WHAT ARE THE AUTHORITIES DOING ABOUT THIS NEW MENACE? NOT MUCH, IT APPEARS!

Such a bizarre mock-up will no doubt backfire on a group who are obviously struggling to get serious backing.

It seems that whilst various councils and the like destroy the countryside of Kent with Euro-tunnels and shopping malls, all the land-owners are interested in is turning a native creature into some kind of barbaric monstrosity. However, the summit only managed to attract one local politician despite many invitations being sent.

Damian Green, Conservative MP for Ashford, commented, *"If the wild boar population is growing very fast and is likely to start intruding on human life then we need to find a method of control*." Views upon such methods were varied at the meeting. Some land owners wanted eradication, some wanted numbers kept down by hunting, but some said they should be treated as an endangered species.

Derek Harman, a boar expert and former gamekeeper, showed various footage of wild boar to the summit participants and then said: *"I disputed a Ministry of Agriculture estimation last year that stated the boar population was only around 100. Since the government figures last year there have been fifty boar shot in the Kent and East Sussex border and that hasn't made any dent in the population.*

I think they should be classed as game animals exactly the same as deer. The deer legislation would be adequate to cope with wild boar."

One farmer at the summit, Ian Douglas, was seconds away from being attacked by a boar, but shot it in time. However, it still had time to chase his friend over the fields. And despite the encounter he remains slightly sympathetic for the animals.

He told me: *"After looking at the boar on the ground after I'd shot it I thought it was scary, but I do have a sneaking regard for them because there are people out there shooting away just so they can say they have bagged a wild boar, all the while the poor young ones are dying and the big old ones are still there getting on with it. They are the dangerous ones and they are the ones producing more and more."*

Farmers are obviously trying to control something before it may become uncontrollable, but all the while only crops are being damaged and not humans. I find it sad that overblown poster campaigns are being used to put the issue on the government agenda.

When conservationists, landowners and agricultural ministers first heard about the boar re-establishing itself, they never would have expected the situation we have today.

However, such an issue seems to have been dug up more by the campaigners rather than by the tusks of the mighty boar.

It must be said, that whilst no casual rambler wants to stumble upon such a beast, especially when fending for its young, there must come a time when Kent's traditional folk come to terms with nature. In my opinion, the Kent wild boar population is not a major worry. What we must realise is that Britain is not heavily populated by a menagerie of savage creatures and it seems that Kentish folk enjoy the power of being in control of the countryside.

When all we have to contend with are foxes, badger and hedgehogs, and still consider these as pests, it is no wonder we are so overblown in regards to the boar. Imagine the fracas and commotion if wolves or big cats were fully re-established.

For the moment, the boar is healthy, but a few years from now it may be on its way out again. The only hope for it is that it breeds rapidly, even producing smaller, blacker species after inter-breeding with escaped Vietnamese pot-bellied pigs. Until then, this once-native beast will have to once again face the enemy that originally drove it out.

EDITOR`S NOTE: The following article first appeared *in The Planet on Sunday* in early 2000. It is included here to compliment the paper by Neil Arnold.

Un-natural History:
The Boar Hunt

by Jonathan Downes

Once upon a time the gentry sallied forth on horseback to hunt the wild boar. They were the fiercest game animals of the British countryside and were responsible for the deaths of many a foolhardy hunter. They were hunted to extinction by the end of the sixteenth century, although many authorities believe that the only animals left alive in Tudor times were actually semi-domesticated creatures kept in deer parks purely for their use as a game animal.

There were sporadic attempts to reintroduce them into the New Forest in the days of Charles I but they were all wiped out in the chaos which followed the English Civil War. Then, over three centuries later, they began to reappear.

The first inklings of a renaissance in the wild boar population came in the mid sixties when isolated specimens were seen in Hampshire and Surrey. Several were shot and at least one other was killed by a passing car. No-one, however, knew where they had come from.

There were isolated reports of wild boar throughout the next twenty years, but it wasn`t until the beginning of the present decade that they began to appear in any great numbers. This time around, however, there was no doubt at all where they had come from.

With the decline in the farming industry at the end of the 1980s many farmers were facing bankruptcy, and in order to keep their land, which had often been in the family for generations, they often had to adopt completely new business plans. In many cases they threw their farms open to the general public, turning them into mini 'theme parks'. Other farmers started small cottage industries, and threw themselves into the marketing of home-made ice cream, fresh vegetables or arts and crafts, and still more farmers decided that the best way to keep both ends of the wolf from meeting at the door was to continue farming animals for food, but to change to the husbandry of more exotic animals that could be sold for a high price as an exotic foodstuff.

The most popular of these new and exotic farm animals were ostriches, bison and wild boar. The first two species proved to be relatively docile in captivity – there were escapes, but on the whole the creatures were recaptured relatively quickly.

Wild boar, however, proved to be just as much of a problem to the 20th Century farmer as they had been to his ancestors, and within a very few years there was a burgeoning population of wild pigs across many parts of the south coast.

In particular the Romney Marsh area of Kent was (according to some reports at least) over-run with the fearsome beasties, and the popular press began to carry scare stories about cars, and even people who had been attacked by fierce brindled wild swine.

It says a lot about the workings of successive British Governments that it was nearly a decade after the Kent wild boar had first been reported that HM Government actually admitted that they had a problem on their hands. Norman Baker, the MP for Lewes tabled a question at Prime Minister's Question Time on the 2nd June 1998. He asked "what estimate he has made of the wild boar population in (a) East Sussex, (b) Kent, (c) Dorset and (e) elsewhere in the United Kingdom; what is his policy in respect of these animals; and if he will make a statement."

The following year MAFF announced that there could be upwards of a hundred and twenty individuals roaming the south east of England.

It is certain that they have established a strong breeding presence among the stockbroker belt of East Sussex and Kent. What is perhaps surprising is that they have even adapted to a semi urban environment - the omnivorous boar finds rich pickings among the well-stocked dustbins of Tunbridge Wells and Sevenoaks, as well as causing severe damage across farmland and woodland.

More ominously, the wild boar carries with it a real threat of conveying disease between herds of outdoor pigs. With Swine Fever rife on the European mainland there is a huge risk if the disease jumps across the channel. MAFF have even had reports of wild boar mating with outdoor sows, giving rise to litters of stripy piglets.

Across the channel in France, boars have become an aggressive and hairy plague. Overprotection, over-feeding and the human exodus to the cities have led to a boom in the porcine population. During 1998, the French Government paid out a record Fr155 million (£15 million) in compensation to farmers and others for the damage caused by wild game - an increase of a third on the previous year, and four times the figure for 1987.

More than 80 per cent of the destruction was caused by wild boars, which can grow to a weight of 270lb and will eat just about anything, from potatoes and drying laundry to mice.

Even in Hong Kong which, when I lived there thirty years ago had wild boar numbers under a dozen, now has a fine and prolific population which seems to grow each year.

If something is not done about the burgeoning numbers of boar in the English countryside they may well start to cause a significant problem.

It has been suggested in many quarters that boar hunting under license be reintroduced as a British countryside pursuit. I don't approve of blood sports, and would also suggest that the prospect of hardy and aggressive wild swine made angry after being wounded by an inept huntsman ain't really the sort of thing that we want wandering around the Home Counties.

There are ways to control the size of an animal population without resorting to legalised brutality.

Wild boar pellets laced with a contraceptive substance would be a cheap and very effective way of controlling numbers. The numbers do, after all, have to be controlled.

However, is there any real need to wipe the creature out for a second time?

Our species is an obnoxiously xenophobic one and seems to have difficulty in grasping that it is perfectly possible for it to share the planet with other creatures. If we are sensible, there is no reason at all why we should not be able to live quite happily alongside the wild boar – if its numbers are kept under control.

I hope that we can. As we enter the twenty first century, Britain is already too tame a place. The addition of a soupcon of danger can do the countryside, and those of us who love it, no harm at all, and may even do it a little good!

Living Dinosaurs:

A case of mistaken identity

by Richard Freeman

"No; a reptile - a dinosaur. Nothing else could have made such a track."
Professor George Edward Challenger.

No group of animals that have ever lived hold us in such a thrall as dinosaurs. They rampage through our childhood imaginings and stalk across the silver screen from the jerky black and white silent, *The Ghost of Slumber Mountain* to the hi-tech realism of *Jurassic Park*. There are dinosaur theme parks, dinosaur toys, dinosaur sweets, dinosaur clothes, and more books have been written about dinosaurs than any other creatures.

The fascination is understandable: humanity has been dominant on earth for less than one million years. Dinosaurs, however, were the unchallenged rulers of this sphere for more than one hundred and twenty million years! The cause of dinosaur extinction has still to be established. Palaeontologists are divided into two camps over this. The first is the "smoking gun theories" - these evoke great global catastrophes to explain `the terrible lizards'' fall from supremacy.

These include massive volcanic activity in Asia, radioactivity from exploding stars, and the ever-popular scenario of the asteroid hitting the earth!

Those in the second camp point out that dinosaur decline was gradual and look to much more reasonable ideas such as climate change, and new diseases crossing newly formed land bridges as sea levels fell with global cooling.

The idea of total dinosaur extinction is totally false. Dinosaurs are around us all the time and in some ways they are just as successful now as they ever were, for birds are, in fact, dinosaurs.

As far back as 1860 when *Archaeopteryx* (then thought to be the first bird) was discovered in the Solenhofen shale of Bavaria, the link between dinosaurs and birds has been known. The specimen bore tailbones, teeth and claws like any small carnivorous dinosaur, but beautifully preserved about it were feathers. Since then, many feathered dinosaurs have been discovered - some pre-dating *Archaeopteryx* by several million years.

All modern birds can literally be considered dinosaurs. Pneumatic bones, erect stance, and skull fenestrations are among their many shared features and, strange as it may seem, *Tyrannosaurs rex* - the largest and savagest predatory dinosaur - has more in common with your pet budgie than it does with *Triceratops*.

When feeding ducks in the park you are feeding dinosaurs, your local pet shop sells dinosaurs for your home and if you watch your Uncle Ted eating chicken you can boast to your friends you have seen a "man eating a dinosaur". However small feathery things that go tweet are not what the word dinosaur summons up for most people.

Are there any non-avian dinosaurs surviving today? If so they would truly make excellent dragons. Are there any giant sauropods or razor toothed coelurosaurs slinking through the jungles or lurking on remote mesas awaiting formal discovery by incredulous scientists? There are those who think the answer is yes!

We begin our dinosaur safari in the cradle of mankind, darkest Africa.

Anyone boarding the Southampton train from Waterloo station on 23rd December 1919 at 11.30 may well have been startled by two outlandish figures. One was a fierce looking hound that seemed more wolf than dog. The other was a tall, weather-beaten man carrying a rifle.

The man was Captain Leicester Stevens, and his dog was Laddie - a wolf/dog hybrid, a barrage dog who had bravely carried messages under heavy fire in W.W.1. His quest was to travel to Central Africa to hunt a surviving "brontosaurus".

His intentions had made national news and ironically, given what we now know about dinosaurs, an old lady from the Wild Birds Protection Association had written to him asking him not to shoot the dinosaur. Sportsmen, hunters, and demobbed soldiers had written too, asking to accompany him. Perhaps unwisely, he elected to go alone save for his dog.

The pair made it to the jungle, but were never seen again – no-one knows what became of them. Without adequate backup that a team expedition would have provided they fell victim to tropical illness and died alone thousands of miles from home. Sadly it seems the report that had inspired their endeavour was a hoax. It appeared on the 17th November 1919 in *The Times*.

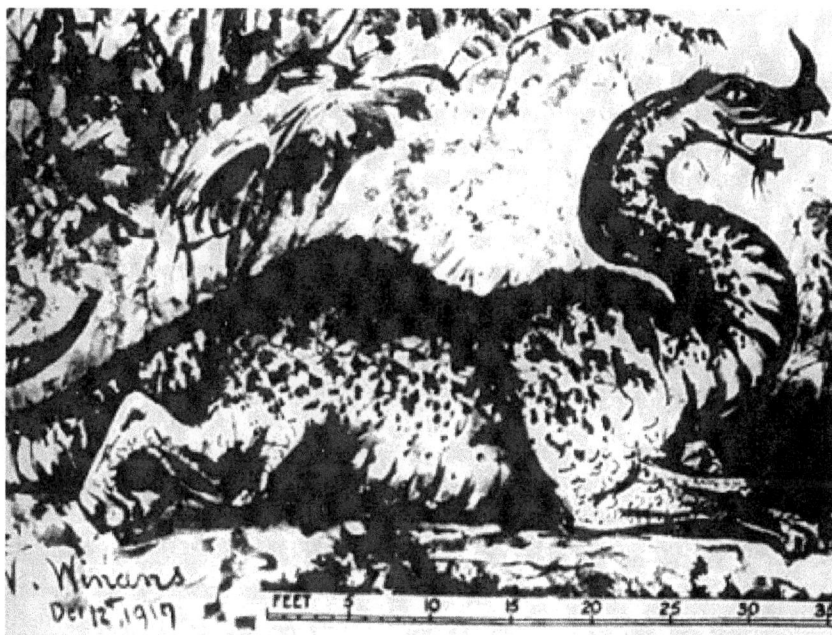

An unlikely dinosaur drawn by Walter Winans for *The Daily Mail* in 1919 to accompany the story of Captain Leicester Stevens' expedition

A TALE FROM AFRICA
Semper aliqud novi

The Central News Port Elizabeth correspondent sends the following:

The head of the local museum here has received information from a M. Lepage, who was in charge of railway construction in the Belgian Congo, of an exciting adventure last month. While Lepage was hunting one day in October he came upon an extraordinary monster, which charged at him. Lepage fired but was forced to flee, with the monster in chase. the animal before long gave up the chase and Lepage was able to examine it through his binoculars. The animal, he says, was about 24 feet in length with a long pointed snout adorned with tusks like horns and a short horn above the nostrils.

The front feet were like those of a horse and the hind hooves were cloven. There was a scaly hump on the monster's shoulders.

The animal later charged through the native village of Fungurume, destroying the huts and killing some native dwellers. A hunt was organised but the government has forbidden the molestation of the animal, on the ground that it is probably a relic of antiquity. There is a wild trackless region in the neighbourhood which contains many swamps and marshes, where, says the head of the museum, it is possible a few primeval monsters may survive.

Firstly the animal described does not resemble a "brontosaurus".

This creature - more properly known as *Apatasaurus* (the name brontosaurus came about due to a mix up in fossil skulls) was a sauropod dinosaur - a long-necked herbivore. The creature Lepage reported more closely tallies with a ceratopsian dinosaur, the group that contained such horned dinosaurs as *Triceratops*, *Styracosaurus*, and *Monoclonius*. Any dino-buff cannot fail to have noticed the glaring errors that make even this identification a non-starter.

Ceratopsians had rounded elephantine feet, not hooves. They also possessed a bony frill about the neck that an observer could not have failed to notice. Horned dinosaurs lacked this odd animal's scaly shoulder hump. Lepage's animal is a complete chimera and the story sounds like complete fabrication.

The unlikely saurian was back in *The Times* on December 4th.

News apparently corroborating the report of the existence in the Congo of a monster known as a Brontosaurus (the thundering saurian) comes from Elisabethville.

A Belgian prospector and big game hunter named M. Gapelle, who has returned from the interior of the Congo, states that he followed up a strange spoor for 12 miles and at length sighted a beast certainly of the rhinoceros order with large scales reaching far down its body. The animal, he says has a very thick kangaroo - like tail, a horn on its snout, and a hump on its back. M.Gapelle fired some shots at the beast, which threw up its head and disappeared back into a swamp.

The American Smithsonian expedition was in search of the monster referred to above when it met with a serious railway accident, in which several persons were killed

Needless to say the Smithsonian did not find this amusing, especially as several of its members had been killed in a railway accident in Africa. This only confirmed the tall tales in the eyes of both the popular press and the general public. The Smithsonian felt it had to quash such outrageous nonsense and wrote a letter *to The Times* published on the 21st February:

Sir, I am authorised to contradict the statement that the members of the Smithsonian African Expedition who proceeded to this territory came here to hunt the brontosaurus. There is no foundation for this statement. I may also state that the report of the brontosaurus arose from a piece of practical joking in the first instance, and, as regards the prospector "Gapelle", this gentleman dose not exist except in the imagination of a second practical joker, who ingeniously coined the name from that of Mr L. Le Page.

Yours faithfully

WENTWORTH. D. GREY

Acting Representative of the Smithsonian African expedition in the Katanga Elizabethville, Jan 21

Another bogus report came in on July 15 1932 in the *Rhodesia Herald* in which a F. Grobler claimed to have knowledge of the existence of a giant lizard known as the *Chepekwe*. Grobler stated that it had been discovered six months earlier by a German scientist in the swamps of Angola. The reptile fed on hippos and rhinos, and Grobler claimed to have seen a photograph of the monster squatting on a hippo it had just killed. Grobler`s *gravitas* seemed supported as he claimed to have acted as a guide to the renowned explorer Hans Schomburgk in his expedition into the Dilolo swamps. The Major had stated in a lecture the previous year that a tradition of giant reptiles was prevalent in Central Africa.

Shortly after this a Swedish man - J.C. Johnson, an overseer in a Belgian rubber plantation - wrote to the *Cologne Gazette* enclosing claimed photographs of the creature. These, together with his story, found their way into the *Rhodesia Herald*. The lurid tale runs thus:

On February 16 last I went on a shooting trip, accompanied by my gun- bearer. I only had a Winchester for small game, not expecting anything big. At 2 p.m. I reached the Kassai valley.

No game was in sight. As we were going down to the water, the boy suddenly called out "elephants". It appeared that two giant bulls were almost hidden by the jungle. About 50 yards away from them I saw something incredible- a monster, about 16 yards in length, with a lizard's head and tail. I closed my eyes and reopened them. There could be no doubt about it, the animal was still there. My boy cowered in the grass whimpering.

I was shaken by hunting-fever. My teeth rattled with fear. Three times I snapped; only one attempt came out well. Suddenly the monster vanished, with a remarkably rapid movement. It took me some time to recover. Alongside me the boy prayed and cried. I lifted him up, pushed him along and made him follow me home. On the way home we had to traverse a big swamp. Progress was slow, for my limbs were still half- paralysed with fear. There in the swamp, the huge lizard appeared once more, tearing lumps from a dead rhino. It was covered in ooze. I was only 25 yards away

It was simply terrifying. The boy had taken French leave, carrying the rifle with him. At first I was careful not to stir, then I thought of my camera. I could plainly hear the crunching of rhino bones in the lizard's mouth. Just as I clicked, it jumped into deep water.

The experience was too much for my nervous system. Completely exhausted, I sank down behind the bush that had given me shelter. Blackness reigned before my eyes. The animal's phenomenally rapid motion was the most awe- inspiring thing I had ever seen.

I must have looked like one demented, when I at last regained camp.

Metcalf, who is boss there, said I approached him, waving the camera about in a silly way and emitting unintelligible sounds. I dare say I did. For eight days I lay in a fever, unconscious nearly all the time".

It seems the herbivorous *Triceratops/Brontosaurus* had been transformed into the savage, carnivorous *Tyrannosaurus rex*. All the more challenging to the intrepid explorer.

Unfortunately Johnson's picture did not live up to his story. It is a tawdry fake showing a Komodo dragon inexpertly superimposed on a dead rhino. So poor is the quality that it would not frighten anyone over the age of 5, let alone send a supposed seasoned hunter into a fear-crazed fever for over a week. (See p.30)

The reader might feel a little disheartened at this point, as all the African stories so far have turned out to be hoaxes. There is however a point to this.

A. We have sorted the chaff from the wheat and can now proceed to genuine reports.
B. The hybrid animal reported seems to have characteristics of both dinosaur-like creatures reported in Central Africa as there are *two* distinct kinds. Moreover, one kind is indeed referred to as *Chepekwe* in some areas. Let us take a look at this beast first.

A dishevelled tramp peddling gridirons wandered up the garden path of Ethelreda Lewis one day in 1925. Most folk would have shooed such an unwholesome fellow off their property but Ms Lewis was a kindly soul and invited him in for refreshments. As it turned out this was a stroke of luck for both the vagrant and Lewis, who was a Johannesburg novelist. The old man began to reminisce about his past and the literary immortality of both himself and his host was assured.

The tatty old gent of the road was one Alfred Aloysius Smith, or Trader Horn as he had been better known. The novelist soon realised she had a gold mine in her living room and transcribed his stories into a series of best selling books.

Horn's tale was the stuff of pulp fiction. He was born in Lancashire and educated in a strict Roman Catholic school (St. Edwards' College, Liverpool). Here he was taught French, Portuguese and Spanish. This did not suit the young tearaway at all and he was soon expelled for excessive wildness and for "always being on the roof!"

He took a ship to the West African country of Gabon and there, aged 17, started work for a British trading company - Hatton and Cooksons - buying ivory and rubber and selling various trade goods. This is where his story really takes off.

Horn claimed all kinds of fantastic adventures, hunting every known jungle beast, canoeing up unexplored rivers, and generally behaving in a manner befitting a character in a *Tarzan* novel. After 5 years of these shenanigans, he came home to Lancashire and married his childhood sweetheart.

Soon after they moved to London and in an attempt to settle down Horn became a reporter, then a policeman. These, not exactly sedate jobs, failed to excite him enough so he joined Buffalo Bill's Wild West Circus and moved to Pittsburgh U.S.A. Here his wife died, and he was gripped by wanderlust once more and glibly shipped his two children back to relatives in England.

What he lacked as a father he made up for as a traveller.

He roamed the world like the Wandering Jew, visiting Mexico, Australia, Madagascar, and - of course - his beloved Africa. Eventually poverty caught up with him and he became a dropout. He ended up in a Johannesburg doss-house. Shortly after he met Lewis, and Dame Fate smiled on him again. So popular was his life story that it was made into a Hollywood film in 1930 (one wonders if Horn ever saw it and if so what he thought!) Horn died, and was buried in Whitstable, Kent in 1931.

The obvious question is how much, if any, of Horn's narrative can be trusted. We must remember he was an old man recalling events of half a century or more ago. Also a warm meal and a roof over his head would have been incentive enough for him to spin the wildest yarns for his host's entertainment. Finally, Lewis herself probably spiced up the stories with a novelist's style.

Perhaps we should not be so quick to reject all of Horn's adventures. Some quite reputable persons have held stock in what he said. One such was Dr Albert Schweizer, who commented *"apart from a few unimportant slips the statements made by Trader Horn about the country are generally accurate."*

It would be surprising if Horn had not heard of "dinosaurs" in Africa and, true to his reputation, he does not disappoint us. Once, by some lakes in the Cameroons, he came across a footprint as large as a frying pan with three toes. This he linked to a creature known as the *Amali* spoken of by pygmy 'bushmen'. He also claimed to have seen carvings of it in their caves. This curious track turns up again in the saga of the Africa monsters.

The great animal collector Carl Hagenbeck, who spent his life studying and capturing animals for the world's zoos (in the days before captive breeding programmes) believed in a giant saurian haunting the swamps of Africa, but appears to have only known of a "brontosaurus" type creature. Some of his informants however also knew of a short-necked, horned beast. Hans Schomburgk, for example, had heard tell of a dangerous animal lurking in Lake Bangweulu in East Africa. The animal was said to kill hippos, but malaria prevented Schomburgk investigating further.

It was another English ex-pat that gathered more information on the horned giant of Lake Bangweulu. J.E.Hughes was born in Derbyshire in 1876, and attended Cambridge. After this, his family apparently expected him to except a career in the Church of England. This apparently repulsed him so much he rebelled much like Trader Horn before him. The British South Africa Company offered him a job as assistant native commissioner in the newly formed civil service of north-east Rhodesia. After 7 years of service Hughes resigned and became a hunter/ trader. He lived for the next 18 years on the Mbawala islands on Lake Bangweulu. He recorded his life in a book, *Eighteen years on Lake Bangweulu*, in which he writes of the monster...

"For many years now there has been a persistent rumour that a huge prehistoric animal was to be found in the waters of our Lake Bangweulu. Certainly the natives talk about such a beast and "Chipekwe" or "Chimpekwe", is the name by which they call it.

I find it is a fact that Herr Hagenbeck sent up an expedition in search of this animal, but none of them ever reached the Luapula or the lake, owing to fever, etc; they had come at the wrong time of year for newcomers.

Mr. H. Croad, the retired magistrate, is inclined to think there is something to the legend. He told me one night, camped at the edge of a very deep small lake, he heard a tremendous splashing during the night, and in the morning found a spoor on the bank - not that of any animal he knew, and he knows them all.

Another bit of evidence about it is the story Kanyeshia, son of Mieri-Mieri, the Waushi Paramount Chief, told me. His grandfather had said that he could remember one of these animals being killed in the Luapula in deep water below the Lubwe.

A good description of the hunt has been handed down by tradition. It took many of their best hunters the whole day spearing it with their "Viwingo" harpoons- the same as they use for the hippo. It is described as having a smooth dark body, without bristles, and armed with a single smooth white horn fixed like the horn of a rhinoceros, but composed of smooth white ivory, very highly polished. It is a pity they did not keep it, as I would have given them anything they liked for it.

I noticed in Carl Hagenbeck's book "Beasts and Men", abridged edition, 1909, p.96, that the Chipekwe has been illustrated in bushman paintings. This is a very interesting point, which seems to confirm the native legend of the existence of such a beast.

Lake Young is named on the map after its discoverer, Mr Robert Young, formerly N.C in charge of Chinsali. The native name of the lake is "Shiwangandu". When exploring this part in the earliest days of the Administration, he took a shot at an object in some floating sudd that looked like a duck ; it dived and went away, leaving a wake like a screw steamer. This lake is drained by the Manshya river, which runs into the Chambezi. The lake itself is just half-way between Mipka and Chinsale Station.

Mr Young told me that the natives once pulled their canoes up the Manshya into this lake. There were a party of men, women, and children out on a hippo-harpooning expedition. The natives claimed that the Guardian Spirit of the lake objected to this and showed his anger by upsetting and destroying all the men and canoes. The women and children who had remained on the shore all saw this take place. Not a single man returned and the women and children returned alone to the Chambezie. He further said that never since has a canoe been seen on Lake Young.. It is true I never saw one there myself. Young thinks the Chipekwe is still surviving there.

Another bit of hearsay evidence was given me by Mr Croad. This was told to him by Mr. R. M. Green, who many years ago built his lonely hermitage on our Lulimala in the Ilala country about 1906. Green said that the natives reported a hippo killed by a Chipekwe in the Lukula- the next river. The throat was torn out.

I have been to the Lukulu many times and explored it from its source via the Lavusi Mountain to were it loses its self in the reeds of the big swamp, without finding the slightest sign of any such survival of prehistoric ages.

When I first heard about this animal, I circulated the news that I would give a reward of either £5 or a bale of cloth in return for any evidence, such as a bone, a horn, a scrap of hide, of a spoor, that such an animal might possibly exist. For about fifteen years I had native buyers traversing every waterway and picking up other skins for me. No trace of the Chipekwe was ever produced; the reward is still unclaimed.

My own theory is that such an animal did really exist, but is now extinct. Probably disappearing when the Luapula cut its way to a lower level- thus reducing the level of the previously existing big lake, which is shown by the pebbled foothills of the far distant mountains."

Perhaps, if we are to believe Mr.Young's tale, the creature's ferocity kept it from being hunted very often. A picture is emerging of a huge dangerous semi-aquatic animal with a single horn and an antipathy towards hippos. Many have come to the conclusion that these creatures are Ceratopsian dinosaurs. These were a sub-order of Ornithischia (bird hipped dinosaurs) and contained such well known horned dinosaurs as *Triceratops*, and *Styracosaurus*. They were all herbivores and were typified by bearing horns and a bony frill - like an Elizabethan ruff - from the rear of the skull to protect the neck. The number of horns varied between species; some

Lake Bangweulu

From Wikipedia, the free encyclopedia

Bangweulu — 'where the water sky meets the sky' - is one of the world's great wetland systems, comprising *Lake Bangweulu*, the *Bangweulu Swamps* and the *Bangweulu Flats* or floodplain. Situated in the upper Congo River basin in Zambia, the Bangweulu system covers an almost completely flat area roughly the size of Connecticut or East Anglia, at an elevation of 1,140 m straddling Zambia's Luapula Province and Northern Province. It is crucial to the economy and biodiversity of northern Zambia, and to the birdlife of a much larger region, and faces environmental stress and conservation issues.

50 km

16
31
Nsombo

17
33
7
30
5
22
6
18
1   2
23
21   20
8
3   4
14
Chambeshi River
Samfya
27
15   24   19
13
9
25
10
21
32
11   12   26
28
29
21
21
Luapula Bridge
Livingstone Memorial
34
Luapula River

Mansa–Samfya–Serenje road
Dirt roads

Satellite photograph of Lake Bangweulu (upper left) and the Bangweulu Swamps (centre)

. Key: 1 Lake Chifunabuli, 2 Lifunge Peninsula, 3 Mbabala Island, 4 Lake Walilupe, 5 Chishi Island, 6 Chilubi Island, 7 Lifunge Mwenzi Island, 8 Nsumbu Island, 9 Lake Kampolombo, 10 Kapata Peninsula, 11 Lake Kang-wena, 12 Lake Chali, 13 Lake Chaya, 14 Lake Wumba, 15 Pook Lagoon, 16 Lupososhi Estuary, 17 Luena Estuary, 18 Lukuto Estuary, 19 Chambeshi Estuary, 20 Luansenshi River, 21 Grassy floodplains, 22 Chichile Island, 23 Kasansa Island, 24 Panyo Island, 25 Nsalushi Island, 26 Ncheta Island, 27 Lunga Bank, 28 Kasenga, 29 Kataba, 30 Lubwe, 31 Kasaba, 32 Twingi, 33 Chaba, 34 Congo Pedicle. *Satellite image credit: Jeff Schmaltz, MODIS Rapid Response Team, NASA/GSFC*

such as *Monoclonius* bore only one horn on the snout.

There are two main stumbling blocks with the dinosaur theory. First and foremost there is no fossil evidence for any species of non-avian dinosaur surviving beyond the Cretaceous period (that ended 65 million years ago). Secondly, there is no indication of *any* species of being aquatic let alone Ceratopsians. So we need to look elsewhere for this beast's identity. Let us examine some more evidence.

The *Daily Mail's* dinosaur fiasco did produce at least one seemingly genuine piece of evidence in what seems to be an honest letter from C.G. James - a gentleman who had resided in Africa for 18 years. His letter was published on December 26, 1919.

"Sir, I should like to record a common native belief in the existence of a creature supposed to inhabit huge swamps on the borders of the Katanga district of the Belgian Congo - the Bang-weulu, Mweru, the Kafue swamps. The detailed descriptions of this creature vary, possibly through exaggerations, but they all agree on the following points:

It is named the Chipekwe; it is of enormous size; it kills hippopotami (there is no evidence to show it eats them, rather the contrary); it inhabits the deep swamps; its spoor (trail) is similar to a hippo's in shape; it is armed with one huge tusk of ivory"

It is useful at this point to realise that Lakes Bangweulu and Mweru are connected via the Luapula river-system (where, supposedly, the specimen was killed).

Identical reports have come in from elsewhere in the Dark Continent. Lucien Blancou, chief game inspector in French Equatorial Africa, collected stories of unknown animals between 1949 and 1953. Some of these seem to refer to an animal like the *Chipekwe*.

"The Africans in the north of the Kelle district, especially the pygmies, know of a forest animal larger than a buffalo, almost as large as an elephant, but which is not a hippopotamus. Its tracks are only seen at long intervals, but they fear it more than any other dangerous animal. The sketch of its footprint which they drew for M. Millet is that of a rhinoceros. On the other hand they do not seem to have said that it has a horn, though they certainly not said that it has not. While M. Millet was at Kelle, in 1950 if I am not mistaken, one of the best known African chiefs in the district came several days march to inform him that "the beast had reappeared". Unfortunately, this is all I can say, for M. Millet left the district in 1951, and I have not been able to go there myself. The rewards in kind which this official offered the pygmies for tangible proof of the animal's presence yielded no result.

Around Ouesso, the natives talk of a big animal which does have a horn on its nose- though I don't know whether it has one of several. They are just as afraid of it as the Kelle people.

Around Epena, Impfondo, and Dongou, the presence of a beast which sometimes disembowels elephants is also known, but it dose not seem to be as prevalent there now as in the preceding districts. A specimen was supposed to have been killed twenty years ago at Dongou, but on the left bank of the Ubangi and in the Belgian Congo"

This report is particularly interesting as the man in question recognised the print as that of a rhinoceros, one of the few animals capable of killing an adult hippo. (The hippopotamus is on of the most dangerous animals in Africa. Despite the cuddly Disney image this animal has, it is in reality totally unpredictable and highly territorial. It also possesses a huge mouth armed with immense curving tusks that can bite a man in two, or rend a boat asunder.) In the Congo, this animal is called *Emela-ntouka,* and this translates as "killer of elephants". Places where both hippos and elephants are scarce or absent are reputed haunts of this aggressive creature who gores the former animals to death with its horn.

Iise von Nolde spent 10 years in eastern Angola and reported, in 1930, events much like the ones related previously. Natives told her of a monster called *Coje ya menia* or "water lion". The name seemed to relate to the roaring sound the animal produced rather than to any resemblance to a lion. She herself heard it's rumbling cry on several occasions. It was said to inhabit water, but was also seen on the bank. In the rainy season, when the Cuanza river was in flood, it moved to smaller rivers and swamps.

One day she met a native in hippopotamus skin sandals. She asked him if he had killed the hippo and he replied that he had found the animal dead, killed by a *Coje ya meina.* On another occasion, a Portuguese lorry driver told of how he had heard of one of these creatures killing a hippo on the previous night. He intrepidly set off to investigate with several native hunters and found the tracks. The hippo's tracks ran for several miles and seemed intermingled with the tracks of its pursuer that none of them could identify. Finally, they came upon an area where the grass and bushes had been smashed and crushed. The mangled cadaver of the hippo lay in the centre of the devastation. It looked as if it had been hacked and ripped by a huge bush knife. None of the carcass had been eaten. It would seem that the only thing capable of inflicting such wounds would be a massive horn.

For me, the clinch in this animal's identity is a photograph taken in 1966 in the Congo by French photographer and naturalist Atelier Yvan Ridel. The photo shows a large three-toed foot print, one of a set that led out of a mass of reeds, up a steep bank, across a small beach and into the river.

The tracks are instantaneously recognisable to any zoologist worth his salt, as the foot prints of a rhinoceros. The nearest rhino populations to the Congo are 1000 miles away in the Cameroons and the Central African Republic. These are black rhino (*Dicerocs bicornis*) the smaller of the two African species and much smaller than the reports of the *Emela-ntouka*. The toes seem a little more elongate than those of other rhinos, and this may be an adaptation to a marshy environment. The rhino's close relatives in the order perissodactyla (odd-toed ungulates) - the tapirs - display slightly elongated toes and are invariably found in swampy biotopes.

The *Emela-ntouka /Chipekwe* is most likely to be not a ceratopsian dinosaur, but a giant semiaquatic rhinoceros. The idea of a water-dwelling rhino may seem strange but the great Indian rhino (*Rhinoceros unicornis*) spends almost as much time in water as a hippopotamus. It feeds mainly on lush water plants such as reeds and water lilies. The Indian rhino also bares only one horn much like the *Emela-ntouka* and unlike the two savannah dwelling African species, who both bear *two* horns.

This unknown species must be a veritable giant. Natives say it rivals the elephant in size. The largest *known* rhino is the African white rhino (*Ceratotherium simum*) that can reach 5 tons in weight and is second only to the elephants as the largest land mammal. A white rhino would have no trouble dispatching a hippo, but if the *Emela-ntouka* does indeed kill elephants it would need to be even more massive.

One prehistoric rhino *Indricotherium* was the largest land mammal of all time reaching 20 tons in weight, bigger than the largest mammoth. One group of rhinos the *Amynodontids* specialised in an aquatic lifestyle. These flourished in the Oligocene epoch 38 to 25 million years ago finally dying out around 10 million years ago. Could one species have survived into the present? This is by no means impossible, but it is perhaps more likely that our unknown giant is a *modern* species that has avoided detection rather than a prehistoric survivor.

But what of the ivory horn? Rhino horn is made from keratin - a fibrous material that also forms human finger nails and very different to ivory. This is the only sticking point with the rhino theory. Could the natives be mistaken on this point? I think the answer is yes. However much we want this creature to be a dinosaur, the bulk of the evidence points towards a giant aquatic rhino.

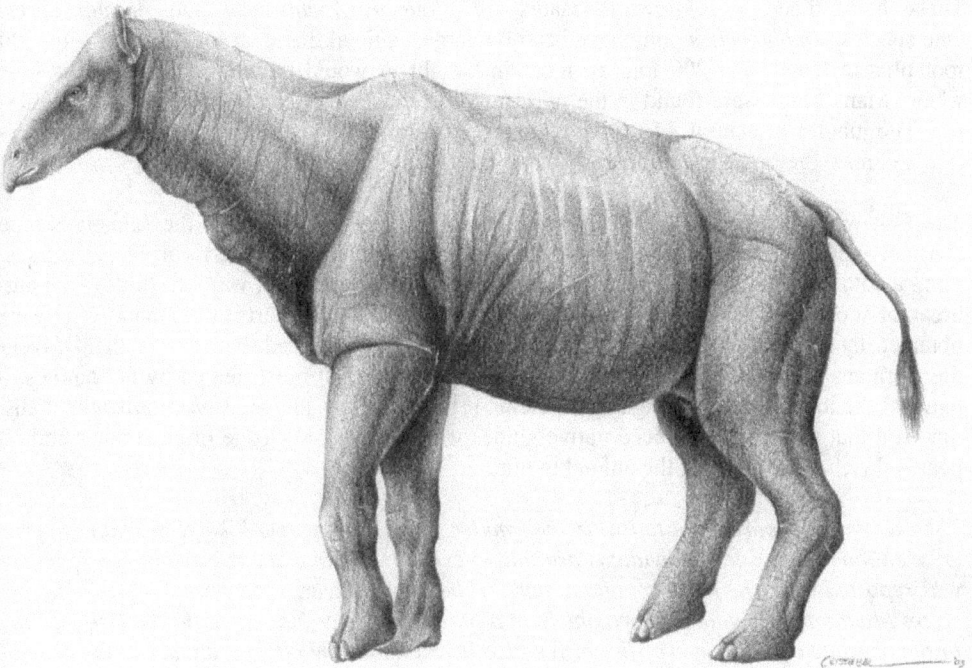

So it seems that our horned "dinosaur" is no such thing, but what of the second possible dinosaur that lurks in the African rain forest? This is a very different beast.

Lewanika was a king who ruled over the remnants of the Barotse Empire on the middle of the Zambezi river. Barotseland lay in the north-western district of what is now Zambia prior to the 1920s. King Lawanika was fascinated by the animals of his kingdom and studied them in detail. His subjects repeatedly told him of a vast aquatic reptile larger than an elephant. The king gave orders to be notified immediately the next time such a creature appeared. The following year three men came to his court and told him they had just seen a monster on the edge of the marshes. They described it as taller than a man with a snake-like head on a long neck. On seeing the men, it slid on its belly into the deep water. The king rode at once to the spot and saw the depression made by the creature, and the channel where it had slid into the swamp. He told a British resident, Colonel Hardinge, that the channel was "*as large as a full sized wagon from which the wheels had been removed*"

The wagons used by the Boers at the time were 1.40 metres wide, so whatever made the track was a substantial animal. Hardinge found out the natives called the creature *Isiququmadevu.*

The description given by the king's three subjects is one we see time and time again across sub-Saharan Africa. An elephant-sized beast with an elongate neck terminating in a small head, a barrel shaped body with four sturdy legs and a long whip like tail. The description is reminiscent of a group of saurischian (lizard hipped), herbivorous dinosaurs called sauropods. These included such well-known dinosaurs as *Diplodocus, Apatasaurus,* and *Brachiosaurus.* One species, *Amphicoelias,* may have been the largest animal that ever lived. At possibly 200 foot plus in length and 200 tons or more in weight, it would have dwarfed even the blue whale. Many kinds were found in the African continent such as *Vulcanodon* and *Aegyptosaurus.* The jungles of central Africa have remained largely unchanged since the Cretaceous period and many believe they still harbour dinosaurs.

The first western involvement with these creatures came in 1913 when the Likuala-Kongo German expedition penetrated the northern Congo. It was led by Freiherr von Stien zu Lausnitz - a colonial officer. The endeavour was due to last two years, but was cut short by the outbreak of World War I. Lausnitz's report was never published, but parts of the manuscript were obtained by pioneering cryptozoologist Willy Ley. Ley discovered that, during their travels, the Germans collected reports of a giant aquatic water animal much feared by the natives. It haunted the lower Ubangi, Sanga and Ikelemba rivers and was known *Mokele-mbembe.* Lausnitz said that several experienced native guides who had no knowledge of each other had repeated the characteristics of the animal to him.

The creature is reported not to live in the smaller rivers like the two Likulalas, and in the rivers mentioned only a few individuals are said to exist. At the time of our expedition a specimen was reported from the none navigable part of the Sanga River, somewhere between the two rivers Mbaio and Pikunda; unfortunately in a part of the river that could not be explored due to the brusque end of our expedition. We also heard about the alleged animal at the Ssombo river. The narratives of the natives result in a general description that runs as follows:

The animal is said to be of a brownish-grey colour with a smooth skin, its size approximating to that of an elephant; at least that of a hippopotamus. It is said to have a long and very flexible neck and only one tooth but a very long one; some say it is a horn. A few spoke about a muscular tail like that of an alligator. Canoes coming near it are said to be doomed; the animal is said to attack the vessels at once and to kill the crews but without eating the bodies. The creature is said to live in caves that have been washed out by the river in the clay of its shores at sharp bends. It is said to climb ashore even at daytime in search of food; its diet is said to be entirely vegetable. This feature disagrees with a possible explanation as a myth. The preferred plant was shown to me; it is a kind of liana with large white blossoms, with a milky sap and apple like fruits.

A sauropod with a horn? One sauropod, *Amargasaurus,* sported a crest of spines on its upper neck. These were believed to have been used for defence against contemporary predators and to have been rattled like porcupine spines for communication and warning. These seem very different to the single horn spoken of here.

The great Belgian cryptozoologist Bernard Heuvelmans has suggested that some confusion exists between the *Mokele-mbembe* and the *Emela-ntouka.* Both are large aquatic animals, both are rarely seen, and now probably exist only in the most remote reaches of the central African rainforest. It seems that some of their characteristics have been transposed upon each other. The horn of the *Emela-ntouka* is erroneously placed on the *Mokele-mbembe* whilst the long tail of the latter is sometimes attached to the former, a feature never found in rhinos. It should be noted that these confusions only occur in a small minority of reports.

Lucien Blancou, whom we met earlier, was also aware of this second kind of water monster. In the 1930 the Linda Banda people described to him an animal they called *Ngakoula-ngou.* It was a gigantic snake-like animal that killed hippos without leaving any sign of a wound, and browsed on trees without leaving the water. "Snake-like" seems to refer to the creatures' necks as no true snakes eat vegetation.

Blancou was told by Yetomane - a chief and hunter of great renown - that in 1928 one of the monsters had crushed a field of manioc belonging to the chief, and left tracks 1.90 metres wide. This probably meant the track left by the animal's body rather than individual footprints. The size is quite comparable to that left in the marshes of Barotseland. The same animal was said to have killed a hippo (what is it with African water monsters and hippos?) in the River Brouchouchou. The villagers ate the corpse.

The Baya people told Blancou of an identical creature they knew as *Badigui.* A man called Moussa related how he had seen the beast in his youth.

When he was about 14 years old and the whites had not yet come (about 1880 I suppose), Moussa was out laying fish-traps with his father in the Kibi stream, which runs into a tributary of the Ouaka called the Gounda in what is now the Bakala district. It was one o'clock in the afternoon in the middle of the rainy season.

Suddenly Moussa saw the badigui eating the large leaves of the roro, a tree which grows in forest galleries. Its head was flat and a bit larger than a python's (Moussa spread his hands and put them together to show me the size). Its neck was as thick as a man's thigh and about 4.50 metres long, much longer than a giraffe's; it had no hair/ but was as smooth as a snake, with similar markings. The underneath of its neck was lighter - also like a snake's. Moussa did not see its body.

His father told him to follow him and run away. The animal gave no cry, but Moussa had heard its cry at other times when he had not seen the beast himself. He did not imitate it for me.

According to him, the old men believed the badigui does not frequent places where you find hippopotami, for it kills them.

Along the Cavally River in Liberia, close to the border with The Ivory Coast, there are similar stories. This river has its genesis in the Niam Mountains in a still unexplored area. At a place known as juju rock, a grey dinosaur-shaped animal is said to live. Natives say it is larger than a crocodile and carnivorous. The diet here seems to be at odds with reports from other areas of Africa, but as we shall see later there may be an explanation for this. To my knowledge, this creature has never been investigated by western science, and due to Liberia's current unstable climate it will probably remain so.

Jorgen Birket-Smith, of the Institute of Comparative Anatomy at the University of Copenhagen, was resident in the French Cameroons during the winter of 1949-1950. He was based at Case du Nyong along the river Nyong. He was told by two old hunters of a large animal that inhabited the river Sanaga. He was told that it was larger than a crocodile or hippo and had a long neck like a giraffe. Smith, who had heard rumours of African dinosaurs, drew a picture of an *Apatasaurus*, and the natives instantly identified it as the creature.

The animal was known as the *New*. It was very rare, but one man recalled that as a boy in the 1920s one had been caught near his village. It was in-between a hippo and elephant in size. The village of 200 people were fed by its meat for a whole week. The *Nwe* was said to browse trees overhanging the water and hardly ever came on to dry land.

James Powell, member of the crocodile specialist group of the International Union for Conservation of Nature and Natural Resources, was studying crocodiles in Gabon along the Ogowe and N'Gounie rivers when he heard stories of a dinosaur-like animal. The Fang people, former cannibals who had been gradually migrating towards Africa's west coast from inland for some 200 years, told of a monster they named *N'yamala*.

Powell befriended a Swiss dentist from the Albert Schweitzer Hospital who had married a Fang girl. He accompanied the dentist up the Ogowe to his wife's village. There he became acquainted with the village witch doctor, Michel Obiang, and found the septuagenarian to be highly intelligent. Powell showed the old man pictures of Gabonese animals such as leopards, gorillas, crocodiles and hippos. The witch doctor identified them all. Powell then showed him a picture of a bear, an animal not native to sub-Saharan Africa. This was unknown to him. Then he was presented with a picture of *Diplodocus,* a sauropod dinosaur. He answered, matter of factly *"N'yamala"*. He added that this animal fed on jungle chocolate - a plant with nut-like fruits that grows near river banks and lakes. This recalls Lausnit's description of the *Mokele-mbembe's* food source half a century earlier.

The witch doctor was insistent that the *N'yamala* had no horn and rejected pictures of other dinosaurs as unknown to him. A *Pterodactyl* was not unreasonably identified as a bat.

The following day Powell travelled 80 miles downstream to study the narrow snouted crocodile (*Crocodylus cataphractus*). He repeated his experiment with the population of a small village. The results were the same - *Diplodocus* was instantly identified as *N'yamala*. The villagers told him that it lived in remote lakes in the jungle. None of them had seen it personally. The area was sparsely populated and according to the American Embassy, Gabon is still 80 per cent unexplored.

Powell made a return to the witch doctor's village at a later date and talked to him again. This time Obiang told him of his own encounter with the *N'yamala*. In 1946 he had been half way up the river N'Gounie were the Ikoy tributary branches off. He had camped for several days by a small cove off the main river where the waters were deeper. He observed the monster as it left the water at around 5:00 a.m. and fed on jungle chocolate. It was around 10 metres (33 feet) long and as heavy as an elephant. Obiang said it was as strong as one of the caterpillar tractors used at the construction of the hospital. It had thread-like filaments running down its neck and two "pouches" in the vicinity of the front legs.

These, Obiang stated, were used for storing food, much like a hamster's cheek pouches. This may seem odd, but sauropod dinosaurs possessed a crop much the same as that which birds have. Because they only had small peg like teeth for nipping off vegetation which were totally inadequate for mastication, they processed their food with the crop. This was a highly muscular section of the throat where fibrous plant matter was crushed into a digestible pulp. Sauropods swallowed stones known as gastroliths for this purpose and stored them in the crop for grinding food (the dinosaur equivalent of dentures!). These stones can still be found today, recognisable due to their highly polished nature.

Powell asked Obiang to take him to the spot where his sighting occurred. The witch doctor obliged and Powell discovered a remote lake in the dense jungle swarming with ants and flies. The lake was some 30 metres (100 feet) across by 5.5 metres (18 feet) deep. Obiang was very scared that the *N'yamala* might still be in the area though nothing was seen. When asked if any hunting trophy such as a skull, bones, or skin were ever preserved Obiang replied *"Oh no, no the n'yamala is king of the waters. It never dies. No one ever kills a n'yamala."*

Obiang suggested that to see a specimen Powell should travel down the Ogowe towards the coast. On an island in the middle of a wide part of the river a *N'yamala,* had killed a hippo but lack of time prevented this.

Once home, Powell contacted Paul W. Richards - an authority on the African rainforest. Richards identified the "jungle chocolate" as a species of Landolphia a large group belonging to the dogbane, or Apocyanaceae, that includes vines and lianas.

In January 1980, Powell - together with Dr. Roy P. Mackal, a biochemist from the University of Chicago and vice president of the International Society for Cryptozoology - mounted an expedition to look for surviving dinosaurs in the Congo. Early into their trip they met an eyewitness at a mission. Firman Mosomele claimed to have seen the *Mokele-mbembe* some 45 years ago at a bend in the Likouala-aux-Herbes River just below the town of Epena. He saw a snake-like head supported on a 3 metre (10 foot) neck break the water. Terrified, he paddled his canoe away as the beast's 2 metre (7 foot) back surfaced. The animal was reddish brown in colour. Mosomele said that in the Epena district natives were afraid to go to the riverbank in late afternoon, as this was when the creature came ashore to feed.

The dread that this animal installs in natives was illustrated by the beliefs attached to the *Mokele-mbembe* by locals. Marien Ikole, a pygmy from the village of Minganga, related some of these to Mackal and Powell. If you tell anyone of a sighting, or even talk of the beast, you will die, and this obviously hampers investigation and implies that the vast majority of sightings go unreported. When the monster appears it causes a miniature tidal wave that will wash on to the bank and then suck you back into the water to be drowned. The *Mokele-mbembe* is so huge it can bridge the river. Once, in a time of war, a tribe escaped from their foes by running across the creature's neck, back, and tail as it positioned itself across the river. These beliefs have analogues in many ancient dragon legends around the world and argue for the existence of actual creatures rather than mere myths.

Pascal Moteka, another pygmy who lived near lake Tele (infamous for sightings), recounted to the expedition of a killing of a *Mokele-mbembe* shortly before his birth (around 1950). The fishermen were too afraid of going out onto Lake Tele due to the monsters who entered it via waterways, or *molibos*. The tribesmen cut down some trees of about 15 centimetres (6 inches) in diameter and trimmed off the branches. Then they sharpened one end of each and rammed the blunt ends into the mud at the bottom of one of the waterways to form a barrier against the monster. One of the creatures tried to smash it's way through and whilst entangled on the spikes, the pygmies managed to spear it to death.

There was a great celebration and the animal's carcass was butchered and eaten. However, all who ate the flesh of the *Mokele-mbembe* were poisoned and died soon afterwards. This recalls the almost universal belief in the toxicity of dragon blood. In many British dragon legends so much as a drop may be lethal and many victorious heroes met their end by spilling the blood of their terrible foes. This tale was later confirmed by other fishermen in the area.

Pascal had seen the animals himself, mainly in mid-morning. He related seeing their long necks rise 2 metres (7 feet) from the water, and - on occasion - a rounded back surfacing like a buoy. His intense fear of the creatures prevented him from approaching them and consequently he only observed them from a distance.

Other witnesses saw the animal at much closer range. Nicolas Mondogo said his father had seen a massive animal with a long neck come out of the river and on to a sandbank. It left dinner plate sized tracks and a great furrow where its tail had dragged in the wet sand. This occurred between the villages of Mokengi and Bandeko on the upper Likouala-aux-Herbes River. Close to this spot Nicolas had his own sighting when aged 17. It was 7 a.m. and he was on his way to a Catholic mission at Bandeko. He had paused to hunt some monkeys when a huge animal rose from the river. The water in the area was only 1 metre (3.5 feet) deep so he could observe the underbelly and legs of the animal.

The beast stayed in view for 3 minutes. It had a long neck as thick as a man's thigh, a head that bore a comb like a rooster, and was reddish brown in colour. It was a mere 12 metres (40 feet) away and was some 10 metres (30 feet) long. It stood 2 metres (7 feet) high and possessed a neck of a similar length, giving a height of approximately 4 metres (14 feet), quite comparable to a giraffe. The tail seemed longer than the neck.

David Mambamlo, a schoolteacher, saw it even closer. Only three years before, he had been canoeing just upstream from Epena at about 3 p.m. when a 2 metre (7 foot) head and neck broke the surface only 10 metres (30 feet) from his vessel. He was mesmerised with horror as it rose further out of the river exposing its upper breast. The monster was grey in colour with no visible scales. David picked out a picture of an *Apatasaurus* from a book and said that the animal most resembled that. He subsequently showed the expedition the location of his sighting, a cave in the riverbank 1 kilometre (0.5 miles) from the village. The water level had dropped revealing the cave, but no occupant was spied.

Daniel Omoa, a Ministry of Agriculture worker, said that in July of 1979 a *Mokele-mbembe* had taken up residence in a pool close to the river. This was just north of Dzeke 80 kilometres (50 miles) downstream from Epena. The people saw it leave the jungle and enter the river by a sand bar that had become a small island during the dry season. Elephant-sized tracks were found on the island and a pathway of crushed grass 2 metres (7 feet) wide was found at the river bank.

Mackal and Powell had to return to the States soon after. Though they had not seen the *Mokele-mbembe* both were now convinced of its existence as a rare, but real biological entity as opposed to just a native myth.

In October 1981, Mackal returned to the Congo this time accompanied by Richard Greenwell from the University of Arizona. Greenwell was also the secretary of the International Society of Cryptozoology. This time they came closer than ever to seeing the *Mokele-mbembe.*

Whilst travelling upstream from the village of Itanga they came to an area were the trees briefly thinned out along the riverbank and were replaced by elephant grass. As they rounded a sharp bend something dived off the 1.5 metre (5 foot) bank and into the water with a great splash. Whatever it was, it was large as it caused a wash to flow over the expedition's dugouts. Due to the shaded nature of the area, none of them got a good look at the creature. The pygmies were in no doubt of the creature's identity and screamed out "*Mokele-mbembe!*" in terror. The group searched the area for half an hour but the monster (if that was what it was) did not reappear. The water into which it dived was found to be 7 metres (23 feet) deep.

They travelled to Dzeke, but found that the animal dwelling in the nearby pool had departed some 16 months earlier, in June of 1980. A villager called Appolonair related his sighting of the animal. Whilst hunting monkeys he saw its long head and neck rear up into view and feed on malombo (another name for jungle chocolate).

They were also told that the creatures were once common before the coming of the white man. When motorboats began to come up the river they retreated to more remote areas.

At the village of Bozenzo they were told of a *Mokele-mbembe* that reached with its neck from the river and seized goats. The creature then devoured them. This seems at odds with the insistence that the animal is an herbivore elsewhere. Later I shall reveal why I think this case is genuine. Village men plucked up the courage to tackle the brute after several livestock losses. They attached ropes to their spear shafts and harpooned the monster. It was cut up and eaten. After this the village became cursed. Houses burnt down for no apparent reason, illnesses broke out, and there were strange deaths. As a result the *Mokele-mbembe* became a venerated animal. This occurred in around 1908.

Perhaps this animal has some sort of toxin in its flesh. This is far from impossible. Poisonous birds have recently been discovered in New Guinea and South America's poison dart frogs are some of the most venomous animals known to man. Alternatively, food poisoning could arise if the meat of this unknown animal was not properly cooked. Either way, an outbreak of illness and death would be attributed to a magick curse from the creature.

At Dzeke they found a track smashed through the jungle at a height of 2 metres (7 feet) and a trail of 25 centimetre (1 foot) prints. The village folk still feared the pool where the monster had laired and would not approach it. Had they investigated the pool upon the first expedition they may well have found their quarry.

Since then there have been many attempts to find the *Mokele-mbembe*. An engineer from California called Herman Regusters led an expedition into the Congo with his wife in September 1981.

Originally Regusters had planned to co-run Mackal's second venture but then changed his mind without adequately explaining why. He claimed he saw and photographed a *Mokele-mbebme,* but his shots were hopelessly under-exposed and of no value. This sort of bad luck seems to habitually dog cryptozoologists. Much to the delight of sceptics (who are generally armchair zoologists) cameras seem to be absent or to fail whenever an unknown animal appears. The next expedition suffered a similar fate.

A search for the monster by Congolese scientists took place in the spring of 1983. Lake Tele was visited by a group from the Ministry of Water and Forests. It was led by Dr Marcellin Agnagna, a zoologist from the *Parc de Zoologie*. On 1st May, whilst filming monkeys in the forest, Agnagna was approached by an excited local who bade him to come quickly to the lake. Wading into the water he saw the back and neck of a large animal some 700 feet away.

The strange animal turned its head as if it had heard Agnagna approach. As he raised his camera and began filming he realised he could not see any thing through the viewfinder. He had foolishly kept the camera on the "macro" setting. In the time it took him to realise this and switch the setting his film had run out! Despite this mother of all frustrations he watched the animal through the telephoto lens, obtaining a detailed sighting. The head was held around a metre out of the water and turned from side to side as if listening. It was reddish brown with a long slender muzzle and crocodile like oval eyes. Just behind the neck was a black hump. Agnagna was sure the beast was a reptile but not a crocodile, turtle, or python.

He and two (particularly brave) villagers waded further into the lake. The animal submerged but then surfaced again, staying in view for 20 minutes. Agnagna took some shots with a small 35 millimetre camera, but the pictures were too indistinct for identification.

British explorer Bill Gibbons led two expeditions. *Operation Congo* took place in 1986 and was an Anglo-Congolese effort. Despite being hampered by bureaucratic problems from the Congolese side, they rediscovered Mackal's *Mokele-mbembe* tracks from his second trip thus independently verifying them. They were also told that Lake Tele itself is not the home of the monsters, but three sacred lakes in the surrounding jungle were.

These lakes are holy to the Boha people who claim ownership of Lake Tele. Two expedition members were allowed to spend a day at one of these lakes, never before seen by white men. They were informed that the *Mokele-mbembes* use Lake Tele as a feeding ground and for transit, but that they actually reside in the smaller bodies of water. No evidence for their presence was uncovered however.

Gibbons met up with an American journalist Rory Nugent at Epena. Nugent was conducting a solo expedition by boat, plane, and on foot. At Lake Tele he saw a black periscope shaped object break the surface. It was some 1000 metres away and beyond the range of his camera. He tried to photograph the object, however, but once again the pictures were far too blurred.

The second Operation Congo expedition took place in late 1992. Though no further evidence was gathered, two unexplored lakes were visited - Lake Tibeke and Lake Fouloukou.

At the time of writing two new expeditions are being planned. Hopefully one of them will have the luck to prove conclusively just what this most enduring of "neo-dinosaurs" actually is.

The *Mokele-mbembe* probably exists, the native accounts of it are very consistent, but is it a dinosaur? The answer is almost certainly no. The two major factors against this idea are the same as those that applied to the *Emela-ntouka*. There are no fossils that suggest any kind of non-avian dinosaur survived the extinctions at the end of the Cretaceous period. Other "prehistoric survivors" such as the okapi - a short-necked giraffe from the Central Africa - and the coelacanth - a primitive fish of the "extinct" order Crossopterygii - have fossil precedents for their survival.

Secondly, and perhaps more importantly, sauropod dinosaurs (or *any* dinosaurs for that matter) were not aquatic. The notion of aquatic dinosaurs had it's sorry genesis with Victorian scientists who perceived these giant animals as being too heavy for their legs to support them for very long on dry land. They envisioned sauropods as spending 90 per cent of their lives in water, buoyed up to relieve them of their vast weight and feeding on soft water plants. Only when they needed to lay eggs would they leave this safe environment (it was once believed equally as wrongly that carnivorous dinosaurs could not swim! In fact they were the best swimming of all dinosaurs).

This ridiculous theory has long been disproved.

Sauropod bodies were adapted for life on land with thick pillar-like legs to support them and hollow vertebrae to lighten their body weight. Their lifestyle could be compared to that of modern elephants or giraffes, huge herd dwelling herbivores that browse vegetation. Fossil track ways show sauropods to have been social animals and that the young travelled with the herd and were not abandoned in lakeside nests as once thought. Lastly it was physically impossible for sauropods to have been aquatic on account of their long necks.

Previously they were pictured as strolling along lakebeds with only their heads above the water. Some species like *Brachiosaurus* had their nostrils on top of the head on an elevated crest. Again this was thought to be an adaptation to aquatic living, a type of "snorkel". However the water pressure on a sauropod's neck would have caused it to cave in and if it tried to take a breath it's lungs would have collapsed. Obviously sauropods entered water from time to time as do modern elephants, in order to bathe, cool off, and rid themselves of parasites, but they were not habitual water dwellers. The *Mokele-mbembe* does seem to be genuinely aquatic, and hence is unlikely to be a dinosaur.

If it is not a dinosaur, then what is the *Mokele-mbembe*? The creature that most fits the picture is a giant monitor lizard. It has been suggested that an Australian species. *Megalania prisca,* may have survived to the present day. If giant monitors can exist in the antipodes why not elsewhere? The African animal seems radically different to *Megalania,* and this may be due to it's aquatic habitat. Many monitors favour a semi-aquatic existence. Most of these amphibious species have elongated necks such as the water monitor (*Varanus salvator)*. Monitors are primarily carnivores, but are very adaptable. It is not out of the question that an omnivorous variety could have evolved. One must remember the stories of the *Mokele-mbembe* eating goats - this lends weight to the omnivore hypothesis.

Dr. Ralph Molnar, an expert on Australian fossil monitors, believes some of them were aquatic and bore crests on their heads. This instantly brings to mind the native descriptions of the *Mokele-mbembe.* A 10 metre (30 foot) aquatic, omnivorous monitor seems much more likely than a sauropod species surviving for 65 million years.

But the mighty Congo has not yet given up all of its monsters. After giant crocodiles, unknown crocodiles, and giant monitors it still has another surprise. Whilst showing dinosaur pictures to natives on his first expedition, Roy Mackal was astounded when one girl picked out a *Stegosaurus* as an animal known to her. These herbivorous *Ornithischian* (bird hipped) dinosaurs were typified by a double row of diamond shaped plates that ran in alternating rows along the back of the animal.

At first thought to be defensive spikes, we now know these highly vasculated structures were thermoregulatory devices that helped the creature control it's body temperature. When wanting to raise the temperature of its body a *Stegosaurus* would present it's flank to the sun. This would warm up the blood as it flowed through the wide plates. If, alternatively, it wished to cool down it would turn away from the sun's rays. An elephant's ears serve a similar purpose today. Defence was, in fact, catered for by sharp spines on the powerful tail, and these differed in number from species to species. They flourished during the mid-to-late Jurassic period then fell into sharp decline when replaced by other families of herbivorous dinosaurs in the Cretaceous. They lingered on in the Indian sub-continent until the late Cretaceous.

The girl, Odette Gesognet, was in her early 20s and lived in the village of Bounila. She related that her ancestors knew the animal and she had been warned to hide behind a tree if she ever saw one. The animal had "planks" growing out of it's back. These were festooned with algae as it spent most of its time in the water. The animal was called *Mbielu-mbielu-mbielu.*

Her story was backed up by Bruno Antoine, a 65-year-old man who had worked for 35 years in the French administration. Antoine had been awarded Congolese Medal of Honour by former president Marien Ngouabi. He told the expedition that one of these creatures had been seen at Ebolo some 200 metres upstream from Epena. It was mainly observed just before dusk at around 4.00pm to 6.00 p.m. It was coated with weeds and was fully observable when it left the water.

The reader will now be expecting me to stress how there were no aquatic stegosaurs and how no non-avian dinosaurs survived the end of the Cretaceous period! That said, we cannot glibly ignore these reports.

The people were obviously likening a *Stegosaurus* to some animal they knew. My guess is that if this creature exists it is a giant freshwater turtle with a spined shell. In the Pleistocene epoch a giant spined tortoise called *Meiolania* existed in Australia.

Today some chelonians such as the Bornean hillstar tortoise sport spines. A giant spiny terrapin lurking in the Congo must surely be possible especially when compared to some of it's more outlandish neighbours!

The African "dinosaurs" seem to be a mixed bag of spectacular animals, none of which are in reality dinosaurs. A mixture of ignorance and wishful thinking has transposed the mantle of "dinosaur" upon them but does this hold for the rest of the world?

Let us now move away from the Dark Continent and widen our search to other continents and other creatures. South America must surely show some promise, after all it was in Venezuela that Conan Doyle set his novel "The Lost World". In the dim primal Amazon jungles there must be room for archaic saurians?

In fact the neotropics have produced far fewer "neo-dinosaurs" than the old world. What meagre tales there are I will recount here.

The most dramatic of these adventures was alleged to have taken place in October of 1907. A German called Franz Herrmann Schmidt and his companion Captain Rudolph Pfleng had travelled up the Solimoes River in Colombia. By the twelfth day of the expedition they had reached a remote valley with a plant choked, shallow lake. Fed by hot springs the vegetation was exceedingly luxuriant, though both men notice a lack of fauna.

Some massive tracks were spied leaving, then returning, to the lake. There were three sets, one adult and two apparently youngsters. Nearby trees had been browsed to a height of around 14 feet.

Next day, whilst canoeing down the lake, a splashing and crashing was heard from the vegetation on the bank. Monkeys in the trees went wild and a huge shape loomed from the shadows. The Indian guides panicked and paddled further out into the lake, then....

"The head appeared over bushes ten feet tall. It was about the size of a beer keg and was shaped like a tapir, as if the snout was used for pulling things or taking hold of them. The eyes were small and dull and set in like those of an alligator. Despite the half dried mud we could see that the neck, which was very snake like, only thicker in proportion, was rough knotted like an alligator's side rather than his back.

Evidently the animal saw nothing odd in us, if he noticed us, and advanced until he was not more than one hundred and fifty feet away. We could see part of the body, which I should judge to have been eight or nine feet thick at the shoulders, if that word can be used, since there were no forelegs, only some great heavy clawed flipper. The surface was like that of the neck. For a wonder the Indians did not bolt, but they seemed fascinated.

As far as I was concerned, I would have waited a little longer, but Pfleng threw up his rifle and let drive at the head. I am sure that he struck between the eyes and that the bullet must have struck something bony, horny, or very tough, for it cut twigs from a tree higher up and further on after it glanced. I shot as Pfleng shot again and aimed for the base of the neck.

The animal had remained perfectly still till now. It dropped its nose to the spot at which I had aimed and seemed to bite at it, but there was not blood or any sign of real hurt. As quickly as we could fire we pumped seven shots into it, and I believe all struck. They seemed to annoy the creature but not to work any injury. Suddenly it plunged forward in a silly clumsy faction. The Indians nearly upset the dugout getting away, and both Pfleng and I missed the sight as it entered the water. I was anxious to see its hind legs, if it had any. I looked again only in time to see the last of it leave the land- a heavy blunt tail with rough horny lumps. The head was still visible though the body was hidden by the splash. From the instant's opportunity I should say that the creature was thirty- five feet long, with at least twelve of this devoted to head and neck."

The monster dived under the dugout and resurfaced about an eighth of a mile away. The creature began swimming towards them and on seeing the impotency of their bullets the men paddled frantically away, finally losing their giant pursuer behind an island.

This story appeared in the *New York Herald* on Sunday 11th January 1911. Pflend is said to have died of fever in March 1908 so the tale was never backed up by a second testimony. The monster itself is a hopeless hybrid of dinosaur, turtle, and tapir. It sounds like a complete construct. It is interesting to note, however, that at one time sauropod dinosaurs were believed to have trunks much like tapirs. The nasal openings in sauropod skulls are placed high up between the creature's eyes (or on top of the head in some species). This is much like the placement of the nasal openings on elephant skulls. This led some to conclude that sauropods had small trunks like the beast previously described. However sauropod nasal openings lack the tiny diagnostic scars left by muscle attachments as seen in elephants. This theory has therefore largely been abandoned.

It also goes without saying that an animal with the flipper-like limbs reported by Schmidt and co could not have left the elephant-like tracks they also reported.

In 1931, Harald Westin, a Swedish explorer, was travelling down the Marmore River in the Mato Grosso region of Brazil when he happened across an odd animal sauntering along the bank. The 20 foot beast had an alligator-like head and possessed a body resembling a boa constrictor, but with four lizard-like legs. Westin kept up the tradition and took a pot shot at the animal. It made a clucking noise, but seemed unharmed.

The description of this creature ill-fits any dinosaur. Apart from the elongate body this thing (if it existed at all) sounds more like a crocodilian than anything else.

Sebastian Bastos was a guide of ancient jungles educated in Switzerland. In 1975 he met an enormous businessman from Geneva with whom he became friends. Bastos confided many things in his friend including an alleged encounter with a supposed dinosaur.

In the early seventies, he had been canoeing in the rainforest whilst the water levels were unusually low. He was proceeding on foot pulling his vessel behind him towards a native friend whom he had arranged to meet at this particular point. Disturbed by a noise behind him, Bastos looked back and was appalled to see an immense monster rear up from the river and smash his craft like matchwood. He and the Indian ran away in terror pausing only to see if the horror still pursued them. To their intense relief, the brute had dived beneath the surface again. The Indian explained that the natives went in great fear of these animals. They laired in deep holes in the river, in the jungle's interior. At night they would sometimes come ashore. Their heads, necks and backs measured 18 feet (suggesting an overall length of 9 metres - or 30 feet).

This whole story is as suspect as a puppy looking innocent next to a pile of poo! The exact area of the Amazon this adventure is supposed to have occurred in is not identified, the businessman has never given his name, and once again our "dinosaur" is aquatic.

Colonel Percy Fawcett, whom we met with his suspiciously slender giant anaconda in the last chapter, had this to say about dinosaurs in the Amazon....

"... some mysterious and enormous beast has frequently disturbed the swamps- possibly a primeval monster like those reported in other parts of the continent. Certainly tracks have been found belonging to no known animal- huge tracks, far greater than could have been made by any species we know."

I've already made clear my opinions on Percy Fawcett's stories and I have the same feelings on South American "dinosaur" reports.

One can, at a push, conceive of dinosaurs in South America, but in *North* America, in the icy wastes of the Yukon? Just such an outlandish tale surfaced in Paris in 1905. It related the story of Georges Dupy, a French-Canadian missionary and banker, from San Franscisco.

Together with two companions, he was, gold prospecting at Armstrong Creek near the McQuestern River. The men were hunting moose for food when one of the animals they were stalking let out a bellow of alarm and took flight with its fellows. Upon investigation of the area the men discovered the imprint of a huge animal in the snow.

The body seemed to be 30 feet long by 12 feet wide and impressed itself 2 feet deep. There were also four footprints measuring 5 feet by 2 ½ feet. These seemed to bear claws of 1 foot. The animal's tail had also left a furrow some 10 feet long and 18 inches wide.

The group tracked the uncommon spoor for six miles until it ended at the foot of a ravine. They had the impression that the brute had leapt up the side of the cliff. Sometime later a concerted search for the monster was made with the Canadian Mounted Police. One evening, after a fruitless day of trekking across the tundra, they set up camp on the summit of a rock gulch and lit a fire. Dupy took up the story.

"We lay by the fire, relaxed our aching limbs, and let our eyes roam over the marsh, glittering with icicles and hoar frost crystals, that we had just crossed. The tea was steaming ready in the pail when, on a sudden, we were startled by the sound of falling stone tumbling down into the bottom of the ravine, followed by larger boulders.

Then came a harsh appalling roar.

We sprang to our feet and I don`t mind saying that my teeth chattered and it was not cold, either! Right across the ravine, on the side opposite to that were we were camped, the boulders were rolling heavily into the bottom, as a gigantic black and hairy animal slowly and heavily ascended the grade.

From the corner of its mouth a blood-stained frothy slime dripped. Its horrid jaws were munching, munching, munching. The priest, the sourdough, and Buttler unconsciously clasped each other by the arms and tried to shout, but could not utter a sound. And well for us that we were stricken dumb! Our Indians crouched on the ground, their faces ashy and their bodies trembling like aspen leaves. They pressed their faces to the ground and shut out the sight. Buttler suddenly got up and tore down the hill."

"Luckily, the monster had not sighted us! He stopped barley 100 paces from us. Then popping his huge belly on a big flat rock, he stood motionless gazing into the glaring eye of the red and setting sun! It was a sight that may not have been unfamiliar to our giant forefathers in a remote age. The monster stood still for 10 minutes, as did we.

He actually swivelled round his huge neck, and still did not see us. I calculated he was about 50 feet long. He had a sort of rhinoceros horn on top of his jaws and his carcass was covered with black stiff bristles like those of a wild boar. The hair was plastered with mud and frozen muck. I`d put his weight at 50 tons."

"As we watched, a sound like the crunching of bones came from his dripping jaws. Then he reared up on his hind legs , emitted a horribly hollow roar, gave a terrific leap, and vanished up the ravine. We made no attempt to follow him"

Dupy and co went to Dawson City and requested a party of 50 armed men and mules from the governor. The governor did not buy the story or oblige in Dupy`s request. The *Dawson City Daily Nugget* had a field day mocking the whole affair. However, the creature was apparently seen again some five years later. The priest from their original encounter, Pere Lavagneux, wrote to Dupy after his return to France.

"Ten of my Indians and myself have seen again that horrible beast of Partridge Creek. It was on Christmas Eve, and the monster was passing like a whirlwind over the frozen surface of the river, breaking off with his hind feet enormous blocks of ice from the frozen surface. His fur was covered with hoar frost and his little eyes - that was why he probably did not see us when we met him, some five years back, when you were here my son - glittered like fire in the dusk. He had in his jaws something that looked to me like a caribou. He moved at a rate of more

than 30 miles an hour. The temperature stood at 45 degrees below zero. At the corner of the cut off, the monster vanished."

"It is evidently the same monster we saw before. Together with the chief Stinehane and his two sons I followed up the trail of the horrid beast. They were exactly like the track you and I and the rest saw when you were here. Then they were embedded in the muck of the moose lick. Eight times on the snow we measured the prints. They were the same and so was the enormous body. Not a 20th of an inch difference! We trailed them to Stewart, fully three miles, when the snow fell and obliterated the tracks."

It is hard to know just were to begin in picking fault with this lot of hot air. Firstly, the description of the monster sounds a little like *Cereatasaurs* save for the size, twice as large as that dinosaur. Secondly, no known dinosaur bore fur (though some had feathers). Thirdly, Dupy's statement inferring "giant" human ancestors living alongside dinosaurs shows his gross ignorance. Fourthly, the fact that the animal had not grown a fraction in five years. Fifthly, the extremely low temperatures cited are not conducive to giant reptiles. The hairy Yukon dinosaur can safely be thrown into the cryptozoological out-tray.

Moving further afield, let us examine the world's greatest continent, Asia. There are many reports of dragon-like creatures here in their legendary home, but few that resemble dinosaurs.

Lake Tian Chi or Heavenly Lake lays in the crater of Baitoushan volcano, in Jilin province, China. In 1980, a group of meteorologists were visiting the area when they encountered a "water dragon". The creature was described as being larger than a cow, with a metre long (3 foot) neck and a flattened duck-like beak. One of the scientists took a shot at the creature. It seemed unharmed and dived back beneath the surface. More sightings were reported in 1994 and film crews from China, Japan, and a team of North Korean scientists visited the remote lake. A photograph was taken, but in true monster snap tradition it was too blurry to be of any use

The North Korean group would have done just as well to stay at home as a near identical monster has been reported in their own country. Chang Bai Tien or Long White Heaven Lake, has a duck-billed denizen. Like it's Chinese counterpart it is remote and rarely visited.

It has been suggested that these animals are surviving hadrosaurs or "duckbilled" dinosaurs.

These were herbivorous dinosaurs with flattened jaws superficially resembling a duck's beak. Unlike ducks their bills were packed with teeth, up to 2000, more than any other dinosaur. These tiny file like teeth were used to shred tough, fibrous food such as pine needles, twigs, and seeds.

They were once believed to be aquatic on account of their bills and a specimen with fossilised skin still attached (this is ultra rare as soft tissue usually decays soon after death). The digits of this individual appeared to be webbed re-enforcing the idea of amphibious hadrosaurs. It was soon found out, however, that in life the skin of the dinosaur's foot bore pads like those of a camel. The skin had shrivelled after death to give the appearance of webbing. This, together

with the dentition adapted for dealing with land plants, pointed to a definite terrestrial lifestyle. Clearly the duckbilled oddities in these lakes cannot be hadrosaurs.

Heading south to the Malayan peninsula we come to Lake Tasek Bera and it's odd inhabitants. The local Semelai people speak of long-necked aquatic animals that dwell here. The animals are harmless to humans and feed only on water plants. The small head at the end of the neck is furnished with two small soft horns.

In the late 50s an officer with the Malayan police force went swimming in the lake. After mooring his boat beside Tanjong Keruing - a small headland - he dived into the water.. Looking back over his shoulder he saw a huge neck rearing up over a clump of rassau weed some 36 metres (120 feet) away. Two silvery grey curves showed behind the neck. Panicking, the man swam back to the boat and paddled away as fast as he could. Looking back one last time, he saw the monster watching his retreat.

The man's commander passed on his account to Stewart Wavell, a producer for Malayan radio, commenting on the man's fine record and reliability. So impressed was Wavell that he travelled to the lake in 1957 in the hope of recording the animal's call said to resemble an elephant's trumpeting.

Wavell made camp with his two guides on Tanjong Keruing. Whilst preparing his wire recorder and camera the monster's cry boomed out across the lake…

"A single staccato cry from the middle of the lake…It was a kind of snorting bellow, shrill,

strident like a ship's horn, an elephant trumpet, and a sea-lion's bark all in one."

He switched the recorder on, but the call did not come again. In 1962, an R.A.F expedition visited the lake but found no monsters.

The aquatic nature of these animals once more rules out dinosaurs. The small soft horns sound remarkably like those of a giraffe. These curious appendages are called ossicones and are possessed only by the giraffe and it's smaller relative the okapi. So, an aquatic giraffe? Maybe not but the beasts of Lake Tasek Bera would seem to be some kind of mammal.

Australia, that lost continent of primitive mammals and giant reptiles, also has it's stories of living dinosaurs and ones much more savage than anywhere else in the world.

In the remote area of northwest Australia called Arnhem Land the aborigines tell of a giant bipedal flesh-eating monster fully 9 metres (30 feet) tall. They call this creature *"Burrunjor"*. The descriptions fit *Tyrannosaurus rex* as no other carnivorous dinosaur grew this large. Even so T.rex is known only from North American strata, its fossils never having been found in Australia.

Such "trivia" is of no interest to those who have claimed an encounter with it!

One such fellow was a policeman who was trailing explorer and bushman Bryan Clark. In the mid 70s whilst mustering cattle in the Urapunji area he lost his way in the bush.

It took him a full three days to find his way out of the wilderness and back to his homestead. Unknown to him a mounted policeman and two aboriginal guides had picked up his trail and were following him.

On the first night of the search, they camped out on the edge of the *Burrunjor's* territory despite the frightened protests of the guides. In the middle of the night the policeman awoke to find his companions screaming in a state of panic and madly attempting to pack their saddle-bags. The ground was shaking as if monstrous foot falls approached, and the snorting of a huge animal was heard. The policeman made like his somewhat more alert colleagues and beat a hasty retreat. He later warned Bryan never to return to the area because if he became lost again he would be on his own - the policeman refused to go near the place, period!

Others claim much closer encounters than this. Between Queensland and Northern Territory in the coastal borderlands is another of the *Burrunjor*. Here in the 50s, cattlemen lost stock to something that left half eaten bulls in it's wake like so many breadcrumbs. It also left what were apparently bipedal reptilian tracks. Gathering a posse, they followed the trail with cattle dogs into swampy scrubland. Suddenly the dogs turned back, and in the distance some of the men were said to have spied a 30 foot tall reptile standing erect amongst the trees. The posse then rapidly lost its enthusiasm.

Another man who claims to have seen one of these brutes is Johnny Mathews, a part-aboriginal tracker. He saw a 25-foot tall reptile in 1961 stalking through scrub near Lagoon

Creek on the gulf coast. *"Hardly anybody outside my own people believes my story, but I know what I saw"* he said to researchers.

Other down under "dinos" are less aggressive. The Central Australian aborigines have a creature in their folklore called *"Kulta"*. According to their traditions *Kulta* lived in swamps and ate only plants. The people feared it because it shook the earth when it walked. It was a quadruped with a long tail and a long neck ending in a small head. It was believed to be so large that if it entered a forest it's head would protrude from one side and it's tail from the other. When deserts overtook the central plains and the swamps dried up *Kulta* died out.

"Lpilya" was a similar creature from the legends of the tribes around the Gulf of Carpentaria. It was said to be 100 yards (300 feet) long and lived in swamps. The tribespeople associated it with thunderstorms on account of it's booming voice

"Wanambi" is another dinosaur-like creature from aboriginal folklore. It was identical to the creatures above except for a colourful crest that ran down its neck and back. Some sauropod dinosaurs did indeed have crests. This was most evident in *Amargasaurus,* a small sauropod we met earlier, with a row of spines along the neck for defence and (via rattling them porcupine-style) communication. Folklore aside, some claim to have seen these latter-day Aussie long-necks.

In the Singleton district of New South Wales stories of the "Dinosaur Swamp Monster" have been circulating for generations. This little known tract of remote swamp stretches beneath the

Blue Mountains. The modern sightings began in 1953 when some duck hunters, Ernie Milling-ton and Horrie Chilvers were out in the swamp. About 30 metres (100 feet) away behind some tall reeds the men noticed a strange animal. Ernie later said…

"All we could see was this long necked beast with a snakelike head, whose neck was thicker than a man's leg, standing about six feet out of the water. We could see the ripples as the ani-mal moved about, but we could not get a look at it's body which was obscured by the tall swamp grass. It moved away to disappear in the grass and swamp scrub. We did not see it again, nor did we want to, and we did not wait around."

Years later the same creature (or at least the same species) was still haunting the swamps. Peter Garland believes what he saw on a remote dirt track in 1981 was a dinosaur. He was attempting to locate a farm recently acquired by a friend and took the wrong turn on a remote road, a turn that led him into Dinosaur Swamp.

"Recent rains had eroded the track, and deciding it was too rough for my new car I stopped, got out, and began walking some distance ahead to see if driving conditions would improve. I wasn't even sure now that I was on the right track to my mate's place."

"Anyway, about this time I had to answer a "call of nature". I walked off the road for a mo-ment. I was about to leave when I looked to one side of me at the sound of rustling shrubbery and stood petrified. There, only yards ahead of me and looking straight at me, was a large, greyish, scaly, reptilian beast something like a brontosaurus approaching me!"

"Coming to my senses, I turned and ran for the car- but not before I noticed it had a large ser-pent like head on the end of a long, thick neck. I couldn't see the end of its tail in the foliage but it must have been long. I reckon the animal was up to 25 feet in length. It stood about 3 feet off the ground on four powerful looking legs."

"The car was about 500 yards away. As I ran towards it, I could see the creature crashing it's way into the scrub in the opposite direction. I lost no time in leaving the place, which I found out later, was not the road to my friend's place at all."

We must remember that these accounts are coming from areas that also have reports of giant monitor lizards. Monitors are known to stand on their hind legs. Monitors can have long necks. In the last chapter we met Megalania prisca the giant monitor said to still exist in the outback. Non-avian dinosaurs died out 65 million years ago, Megalania conversely was around until only ten thousand years ago. Now in the existent today stakes which pony would you back? The so-called dinosaurs of Australia are probably giant monitor lizards.

Finally, for the most outlandish account of living dinosaurs we must move to New Guinea. Explorer Charles Miller and his new bride were honeymooning among cannibals in New Guinea (top marks for originality there, I hope I can marry a girl up for this kind of honey-moon!)

Miller claimed to have uncovered a lost tribe called the Kirrirri . One member of the tribe was using an odd tool to break open coconuts. The object appeared to be some kind of horn of tusk. When enquiring about it's origin to the village elders, Miller uncovered a remarkable story. Wroo, an old man from the village drew a long-tailed, hump-backed lizard in the sand. About it's neck he drew a frill and he adorned it's back with triangular spines. The beast was said to be 12 metres (40 feet) long. The horn had come from just such a brute. The natives called it the Row after it's cry, somewhere between a snake's hiss and a roar. Miller determined to see the animal for himself. The natives told him that the monster's lair lay two or three days to the north-west.

Taking guides he travelled to a triangular marsh in the hill were his native companions became nervous. Then out of a strand of reeds a gigantic animal emerged. Miller's new wife clutched at the grass, paralysed with fear. The following turgid narration is entirely Miller's and I take no responsibility!

"As if in obedience to my wishes, the colossal remnant of the age of dinosaurs stalked across the swamp. Once it's tail lashed out of the grass so far behind it's head I thought it must be another beast. For one brief second I saw the horny point. I heard it hiss- Roooow-Roooow-Roooow."

Miller started to fill the monster but the thing seemed to become alerted to his cameras whirring.

"Suddenly it stopped, reared up on it's hind legs, it's small forearms hanging limp, and shot it's snaky neck in our direction. It was a full quarter of a mile (400 metres) away, it couldn't possibly hear the camera, but I found myself cowering back as if that snapping turtle-shaped beak would lash out and nab me.

I gasped with relief when the creature settled back. Twice more the Row reared up, giving me a good view of the bony flange around it's head and the projecting plates on along it's back bone. Then with a click my camera ran out just as the Row slithered behind a growth of dwarf eucalyptus."

This film, if it exists, has never been shown to zoologists. Miller never produced so much as a still from it in his book "Cannibal Caravan". Moreover as Bernard Heuvelmans, the father of cryptozoology, noted in 1955 the Row is a hopeless hybrid of several totally different dinosaur species. The beak and frill belong to a ceratopsian dinosaur such as Triceratops , the plated back belongs to Stegosaurus, and the elongated neck and tail are those of a sauropod such as Diplodocus. I would add to this the bipedal gait and carnivorous nature of Tyrannosaurus rex or Allosaurus.

The Row is a hopeless hybrid akin to something dreamed up by Toho studios of Godzilla fame and must vie with the hairy Yukon dinosaur for the title of "least likely monster".

The outlook for the survival of non-avian dinosaurs is poor, worse it is non-existent. Not a shred of evidence that these magnificent animals persisted beyond the end of the Cretaceous period 65 million years ago. But what if they had other descendants than birds? Ones that, due to their very nature, could leave no fossil trace. Just such a theory was put forwards by Peter Dickenson in a remarkable book published in 1979.

In The Flight of Dragons, Dickenson attempts something not done since Charles Gould`s Mythical Monsters , to explain fire breathing, winged dragons as real animals within the known zoological frame work.

Impressed by the universality of dragon legends, Dickinson believed they had a basis in fact. The main stumbling block was the sheer size of dragons, animals that after all were supposed to have flown. Looking at mediaeval reconstructions of dragons he reckoned their weight to be around 9,000 kilograms (20,000 pounds).

In order to be able to fly by the muscular power of it's wings a dragon of this weight would need a wingspan of over 180 metres (600 feet) far too massive to be real. And how could an animal possibly breathe out fire? These problems seemed insurmountable until a chance viewing of the crash of the Hidenburg .

"…one day I happened to see on television an old newsreel film of the wreck of the airship Hindenburg, an almost in a flash all my ideas changed. As I watched the monstrous shape crumpling and tumbling in fiery fragments, with the smoke clouds swirling above, I said to myself, it flamed and it fell, and my mind made the leap to Jordanus. All the pieces I had been considering shook themselves into a different shape. I saw that the Hindenburg was not just a very big machine which flew-it was a machine which could fly only because it was very big. Other answers slotted into place.

1. Dragons could fly because most of their bodies were hollow, and filled with a lighter - than - air gas.
2. Dragons needed an enormous body to hold enough gas to provide lift for the total weight of the beast.
3. Dragons did not need enormous wings, because they used them only for propulsion and manoeuvring.
4. Dragons breathed fire because they had to. It was a necessary part of their specialised mode of flight.

Dickenson's theory held that dragons evolved from large fast moving carnivorous dinosaurs like Tyrannosaurus rex. They developed huge chambered stomachs that they filled with hydrogen gas thus achieving flight. The hydrogen was formed from a mixture of hydrochloric acid in the gut and calcium from the bones of their victims and controlled partial digestion of their own bone structure.

The calcium taken from their own bones was being constantly replaced with a regular intake of limestone. This may explain the dragon's legendary love of lairing in caves.

The vast body was filled with this gas and the animal acted, in essence, as a living hot air balloon. As any chemist will know hydrogen mixed with oxygen is highly flammable. This is where the dragon's most famous attribute, its fiery breath, came into play. Dragons needed to breathe fire in order to control their flight. To rise, they filled their gas bag stomachs, and to descend, they burned off gas by breathing it out, possibly with a chemical catalyst, such as fire, much in the same way as a hot air balloon's burner does.

The fiery breath doubtless doubled up as a formidable weapon, a punishing jet of flame with which to destroy prey, and as a display to other members of its species. A similar weapon is employed by the bombardier beetle (Brachinus) that spews a jet of boiling chemicals at its enemies. The two chemical components are produced from different glands and do not reach such high temperatures until it is outside of the beetle's body.

The wings were formed from the extended ribcage much like that of the modern lizard Draco volans , a small gliding species often called the "flying dragon". These were covered with a bat-like membrane and were used in navigating the animal in flight.

Dickenson also believes he can explain some of the more esoteric aspects of dragon legends. The cult of dragon worship would have sprung up from primitive people's fear of such a terrifying creature. The famous "dragon hordes" would have been built from offerings to appease the monsters. Dickenson says dragons would have used gold as a nesting material as it is non-combustible and fairly soft. Their fondness of virgins may have its genesis in human worship and sacrifice of high born victims, perhaps born and raised specifically as sacrifices to dragons.

The theory also provides a good reason why there are no known dragon fossils. In life, a thick mucous lining in the stomach walls kept the powerful hydrochloric acid needed to produce hydrogen in check. After death, the mucous lining was no longer generated and the acid destroyed the animal's body. The creature literally digested itself. Hence, no dragon bones and no dragon fossils. It is for this very reason that Dickenson's theory is impossible to prove. In effect, he is hoisted by his own petard and his wonderful theory must, for the time being, remain just a tantalising possibility.

So far we have considered the possibility that dragons were some kind of physical creature. Perhaps we are barking up the wrong tree. Can such a magickal, powerful beast be confined to mere flesh and bone? Could the origins of dragons lie outside the boundaries of standard zoology, or even cryptozoology?

Humans in Captivity

by

Chris Moiser

Humans as a species fascinate us. There is nothing wrong about that, they must, otherwise there would not be the vast numbers of students studying Anatomy, Anthropology, Sociology, Psychology and Human Biology that there now are. In modern times we temper our curiosity with an understanding and politeness that was not always evident in the past. We still steel a look at individuals who appear different though, whether that difference is racial, through disability, or just because the individual is a prime specimen, (or possibly exceedingly sexually attractive?).

In the past our interest was often less well controlled and more basic, and accordingly was rather exploited. As a result the humans who were exhibited could be seriously exploited.

It is commonly known that slavery was prohibited in Great Britain in 1807, but in fact this was only the slave trade that was abolished, there was not total emancipation in the British colonies until 1833. The start of the legal movement against slavery in Britain had started much earlier with *Somerset V. Stewart* in 1772 when it had been established that the English Courts would not view slavery as an enforceable contractual arrangement, i.e. the slave could not be forcefully sold abroad by his master.

In the following year the Scottish courts ruled that they would not enforce a masters legal claim to a runaway slave. So although the practical controls on slavery started in 1772 it did not free black slaves in England and it was over sixties year later before the last of the British slaves were freed.

In practical terms this means that the earliest of the travelling menageries could have exhibited humans of a non-British origin as caged slaves. We have no direct evidence of this happening but an advert in the *Exeter Flying Post* for 31st July 1788 describing the visit of an unnamed menagerie to Kingsbridge refers to the *"Ethiopian Savage, a surprising Wonder from Botany-Bay, brought over with Capt. Cooke's Curiosities, name unknown, has a striking resemblance of negro men, and eats, drinks, and sleeps in the human way"*.

Cook had in fact returned from Botany Bay on 12th June 1771, some 17 years previously, although "his curiosities" may, of course, have arrived later. We do not know how this individual was accommodated, and whether he was caged as a savage slave, or whether when the show was closed, he worked and travelled alongside the other show staff.

If there was some doubt as to whether the Australian aborigine was exhibited in a cage there is no doubt that Saartjie Baartman was when she appeared in London in 1810. Saartjie (Afrikaans for "little Sarah") was a South African bushman or hottentot woman (there is a difference between bushman and hottentot, but it is often blurred, and remains so in the case of this lady). She came to England at the suggestion of her employers' brother for the specific purpose of exhibition, which she believed would make her rich. Shortly after her arrival she became known as the "Hottentot Venus".

As soon as she arrived in Piccadilly she went on exhibition in a cage on a raised platform. A member of the "African Association", described the cage, and stated that: *"On being ordered by her keeper, she came out... The Hottentot was produced like a wild beast, and ordered to move backwards and forwards and come and go into her cage, more like a bear in a chain than a human being."*

As a result of the exhibition there was a court case, where she gave evidence, in Dutch that she was not under restraint and understood perfectly well that she had been guaranteed half the profits.

The show continued, and after a tour of the English provinces Saartjie went to Paris where she was exhibited by a former animal trainer for 15 months. It was whilst in Paris that she posed nude for scientific paintings.

However instead of returning to Cape Town as a rich woman, she died in Paris in December 1815 and ended up on Cuvier's dissecting table. At this time Bushmen and Hottentots were considered almost as a missing link between human and ape, and therefore their anatomy was of particular interest. In the case of the women the sexual anatomy was also of great interest because this group of people show a trait known as steatopygia, a condition where a large fat reserve is deposited above the buttocks.

Saartjie's dissection provided an opportunity to confirm this structure to be fat, although her disrobing for scientific study had already enabled external observation to conclude that the structure was mainly fat.

Despite the general view about the possible evolutionary position of the bushmen Cuvier, in his report, mentions that Sartjie possessed an impressive memory, spoke Dutch well, could

UNION ROAD, PLYMOUTH.
ADJOINING COOKE'S CIRCUS.

UNDER THE SANCTION OF H.M. GOVERNMENT, AT THE CAPE OF
GOOD HOPE.

ON MONDAY, THE 5TH OF AUGUST, 1850,
And following days.

EXTRAORDINARY EXHIBITION OF

BOSJESMANS,
OR WILD AFRICANS.

Hours of Exhibition, Two, Six, and Eight, p.m.

THIS singular and rapidly decreasing race of the Aborigines
of South Africa first appeared at Exeter Hall, London, to
illustrate a Lecture delivered by Dr Robt. Knox, M.D., F.R.C.S.

The only real Specimens of the Adult Bosjesmans, who have
ever visited Europe.

These extraordinary human beings have been visited by up-
wards of seven hundred of the leading aristocracy, whose
autographs may be seen on application to Mr S. Tyler, amongst
which are the following distinguished personages—His Grace
the Duke of Wellington—the Grand Duke Constantine and
Suite—Prince Louis Napoleon Bonaparte—Lord Stanley—
General Sir Charles Napier, late Governor of the Cape of Good
Hope—Marquis of Douro—Sir Robert Peel—Prince Douro of
Rome—Marquis of Douglas—Earl Craven—Sir William Massey
Stanley—Lady Peel and Daughter—Sir James Graham—
Marquis of Normanby—Sir Watkins Williams Wynne—Lord
Anson and Party—Jenny Lind, &c., too numerous for insertion
here.

A LECTURE

Introductory to each Exhibition, will be delivered by Mr S.
TYLER, upon their Mental, Moral, and Physical Attributes,
likewise the Geographical Position and Features of the Bosjesmans' Country. With a view of making the Exhibition as
instructive as possible, in the course of the evening, the Bosjesmans will exhibit their mode of

TRACKING AND KILLING THEIR GAME,

THEIR METHOD OF

BUSH FIGHTING,

AND CONCLUDE BY DANCING

THEIR SUPERSTITIOUS DANCE,

The whole of which will be superintended, and their clicks
interpreted by their Guardian, Mr S. Tyler.

Mr S. TYLER, Interpreter to the Bosjesmans, begs to inform
the inhabitants of Plymouth that in consequence of the stipu-
lation entered into by George Bishop, with Her Majesty's
Government at the Cape of Good Hope, to return these
singular specimens of the Human Species to their Native
Wilds, they cannot remain much longer in Europe, and consequently but a short time in Plymouth.

Reserved Seats, 1s. Second ditto, 6d.

speak some English and was learning French when she died.

The Wild West shows, although generally associated with "Buffalo Bill" in the 1880s, probably started in 1840, when the American George Catlin arrived with eight tons of Indian artefacts and was quickly joined by a troupe of Ojibbeway Indians. As a result of this he had an extended run in London. Other humans exhibited during this period included Charles S. Stratton, better known as the midget General Tom Thumb, who arrived with Barnum in 1844. Barnum had actually first became famous for exhibiting human freaks before he entered the circus world in 1871 at the age of 60.

1847 saw the arrival of African Bushmen as a family group, they were first exhibited in the Egyptian Hall, where a newspaper reported them to be: " ...*little above the monkey tribe. They are continually crouching, warming themselves by the fire, chatting or growling...They are sullen, silent and savage -mere animals in propensity, and worse than animals in appearance.*"

In 1850 they appeared in Devon. The *Plymouth and Devonport Weekly Journal* of 1st August 1850 reports *an "Extraordinary Exhibition of Bosjesmans, or Wild Africans"*. The exhibition was open three times daily. The exhibition started with a lecture on their moral, mental and physical attributes, and finished with a demonstration of their hunting, fighting and dancing techniques.

The advertisement also states that they appear with the sanction of H.M. Government at the Cape of Good Hope and that there is an agreement to return them to their "Native Wilds" there.

What is also particularly interesting about this group is that the advertisement claims that they first appeared in London to illustrate a lecture delivered by Dr. Robert Knox, M.D., F.R.C.S. Robert Knox was a brilliant doctor and anatomist whose popularity with his students had proved to be his undoing. In order to obtain sufficient corpses to dissect in class, Knox had made the mistake of employing the services of William Burke and William Hare, the body snatchers turned murderers.

After their capture, and the execution of Burke in 1829, Knox was cold shouldered by the medical profession and was never again able to obtain university employment. Thus he was almost pushed into his private lectures, which he gave in London, Edinburgh and Glasgow. He lectured, and published, almost entirely on anatomy and physical anthropology. His book of 1850 was entitled *"The Races of Men"*, and was one of the most extreme racist views published up to that time.

With the exhibition of the Bushmen there had been a degree of respect shown by the exhibitors and an assurance that they would be returning to South Africa as, apparently, free men. A less happy incident, which involved an individual who had been captured and enslaved, was that of Ota Benga, a Bachichiri Pygmy, from what was then the Congo Free State. He appears to have been enslaved when a neighbouring tribe killed his family. They passed him on to another tribe, who in turn delivered him to Samuel Phillips Verner, an American explorer, sometime missionary, who was looking for "pygmies" to be exhibited at the St. Louis Exposition of 1904.

Ota assisted Verner to persuade another group of Africans to go to St. Louis and build an authentic African village and apparently went as a free man. After the exhibition they returned to Africa and Ota and Verner spent 18 months exploring and collecting. When Verner was ready to return to the States Ota asked asked if he could accompany him and Verner permitted him to do so. Unfortunately once in New York the sale of Verner's collection did not go as planned and Verner rapidly became bankrupt. Whilst trying to sort the financial affairs out Verner went to his family in South Carolina, leaving Ota with the American Museum of Natural History. They, apparently with genuine good intentions, passed him on to the Bronx Zoo, believing that he would be more comfortable there. The Bronx Zoo, then under the directorship of the slightly eccentric Dr. William T. Hornaday, made him into the star feature of the exhibition on evolution. Literally he was in the primate house with an orang utan called Dohung as a roommate.

The local Afro-American community were outraged and some of the local churchmen were concerned that his presence would confirm Darwin's theory of evolution. Under threat of legal action, Hornaday had him leave his cage and wander round the park during the day wearing a white suit. At night he returned to the cage. This did not work well - visitors chased round the grounds, tripped him up and generally poked and prodded him. Affairs reached a climax when Benga made himself a bow and arrows and started shooting at the more obnoxious visitors. Several well-intentioned local ministers of the church extricated him from the park and sent him to a Virginian theological seminary. Fulfilling Hornaday's prediction, *("He did not possess the power of learning"),* Benga left the seminary and started work in a tobacco factory. His tale reached a sad conclusion when, in 1916, despondent that he could not return to Africa, he died as a result of a self-inflicted gunshot wound.

There have, in recent times, been human exhibits in various zoos, but typically involving otherwise unemployed thespians, who "clocked off" as the gates closed. These exhibits have been very much an attempt to boost flagging visitor numbers or to illustrate certain environmental issues for educational reasons. Though several such exhibitions did obtain the desired media attention for the organisations concerned none can claim to have seriously upset public sensitivities or to have raised serious questions of exploitation. Neither, apparently, have they excited so much fascination as the displays of the past.

Further Reading

Gould, S.J. *(1985) The Flamingo's Smile* Penguin Books, London.
Skotnes, P. Ed. (1996) *Miscast: Negotiating the Presence of the Bushmen* University of Cape Town Press

Tinamous in Great Britain

by Dr. Karl P. N. Shuker

EDITOR'S NOTE: We are very grateful to Dr Shuker for allowing us to reprint this rare article which originally appeared in the *Avicultural Magazine* #98, 1992.

To aviculturists, tinamous are well-known for being those nondescript, deceptively gallinaceous birds of the Neotropical Region that are in reality most closely related to certain of the giant, flightless ratites. Rather less well-known, conversely, is that at one time they seemed destined to become exotic new members of the English avifauna, as revealed in this article.

Tinamous are among the most perplexing and paradoxical of birds. Comprising some 40-odd species in total, and ranging in size from 8 - 21 inches, they closely parallel the galliform gamebirds in outward morphology, with small head and somewhat long, slender neck, plump body and short tail, sturdy legs, and rounded wings. Admittedly, their bill is generally rather more slender, elongate, and curved at its tip, and the tail is often hidden by an uncommonly pronounced development of the rump feathers, but in overall appearance they could easily be mistaken for a mottle-plumaged guinea-fowl, grouse, or quail (depending upon the tinamou species in question).

Even so, it would seem that their misleadingly gallinaceous morphology is a consequence of convergent evolution (i.e. tinamous filling the ecological niche in South and Central America occupied elsewhere by genuine galliform species, but having arisen from a wholly separate ancestral avian stock), because detailed analyses not only of their skeletal structure but also of their egg-white proteins and (especially) their DNA have all indicated that their nearest relatives are actually the ostrich-like rheas!

Nonetheless, the tinamous are nowadays classed within an entire taxonomic order of their own, *Tinamiformes,* because in spite of their ratite affinities they have a well-developed keel on their breast-bone for the attachment of flight muscles, and are indeed able to fly - though they are not particularly adept aerially, probably due to their notably small heart and lungs, which would seem to be insufficiently robust to power as energy-expensive an activity as flight.

Equally paradoxical is the fact that although their legs are well constructed for running, tinamous are not noticeably successful at this mode of locomotion either, preferring to avoid danger by freezing motionless with head extended, their cryptic colouration affording good camouflage amidst their grassland and forest surroundings.

Their outward appearance is not the only parallel between tinamous and galliform species. On account of the relative ease with which these intriguing birds can be bagged, in their native Neotropical homelands tinamous have always been very popular as gamebirds - a popularity enhanced by the tender and very tasty (if visually odd) nature of their almost transparent flesh. Accordingly, it could only be a matter of time before someone contemplated the idea of introducing one or more species of tinamou into Great Britain as novel additions of our country's list of gamebirds - a list already containing the names of several notable outsiders, including the red-legged partridge (*Alectoris rufa)* and the common pheasant (*Phasianus coichicus).*

The concept of establishing naturalised populations of tinamou in Great Britain was further favoured by the great ease with which these birds can be raised in captivity, enabling stocks for release into the wild to be built up very rapidly.

And so it was that in 1884 the scene was set for the commencement of this intriguing experiment in avian introduction - the brainchild of John Bateman, from Brightlingsea, Essex. The species that he had selected for this was *Rhynchotus rufescens,* the Rufous tinamou - a 16 inch long, grassland-inhabiting form widely distributed in South America with a range extending from Brazil and Bolivia to Paraguay, Uruguay, and Argentina. In April 1883, he had obtained six specimens of this species from a friend, D. Shennan, of Negrete, Brazil, who had brought them to England from the River Plate three months earlier. Bateman maintained them in a low, wire-covered aviary with hay strewn over its floor, sited on one of his homesteads. By June, they had laid 30 eggs, most of which successfully hatched, and half of these survived to adulthood.

In January 1884, naturalist W. B. Tegetmeier paid Bateman a visit, and became very interested in his plans to release tinamous in England. On 23rd February 1884, *The Field* published a report by Tegetmeier regarding this. However, the first release had already occurred (albeit by accident), for during the summer of 1883 a retriever dog had broken through the wire-roof of Bateman's tinamou aviary, resulting in the death of four of the tinamous, and the escape of seven or eight of the others on to Bateman's estate and thence to the Brightlingsea marshes. Only a small number of tinamous had remained in captivity, but these had increased to 13 by the time of Tegetmeier's visit.

As for the escapees, Bateman recognised that they were in grave danger of being bagged by persons shooting in the area (thereby ending any chance that they would succeed in establishing a viable population), and so in a bid to thwart this he issued a handbill, drawing to the attention of local people the basic appearance and habits of tinamous, and his plans for their naturalisation in England. The handbill read:

"The tinamou, or, as it is called by the English settlers on the River Plate, "Big Partridge", is a game bird, sticking almost entirely to the grass land; size, about that of a hen pheasant; colour when roasted, snowy white throughout. When flushed, he rises straight into the air with a jump, about 15 ft., and then flies off steadily for about half a mile; he will not rise more than twice. Mr. Bateman proposes, after crossing his stock with the tinamous in the Zoological Gardens, to turn them out on the Brightlingsea marshes, which are strikingly like the district whence they came, and he hopes that the gentlemen and sportsmen of Essex will give the experiment a chance of succeeding, by sparing this bird for the next few seasons, if they stray, as they are sure to do, into the neighbouring parishes, as they would supply a great sporting

want in the marshland districts."

To supplement his captive stock, following Tegetmeier's visit Bateman obtained three more specimens of Rufous tinamou from his friend Shennan, and also purchased three from London Zoo. In April 1885, he released eleven individuals onto the Brightlingsea marshes; these, together with 14 hatched from eggs, had increased to approximately 50 or 60 birds by September, according to a second, more extensive report by Tegetmeier *(The Field,* 12th September 1885).

Tegetmeier noted that throughout spring and early summer in Brightlingsea and parts of Thorington the Rufous Tinamou's presence there could be readily confirmed by its very distinctive call, described as a musical 'ti-a-u-u-u" in the case of the cock bird, and sounding unexpectedly similar to that of the blackbird *(Turdus merula)*. Illustrating this similarity is an entertaining anecdote contained in a letter to Tegetmeier from Bateman:

"Mr. Bateman, in his letter to me, states: "A passing gipsy bird-fancier hailed my keeper's wife, after listening attentively awhile, with 'That's an uncommon fine blackbird you've got there, missus,' alluding to the note. 'Yes,' she replied. 'Will you take five bob for him, missus?' 'No; I won't.' 'May I have a look?' 'Yes; ye may.' 'Well I'm blowed!"'

As he well might be, seeing what he regarded as the note of a blackbird proceeding from a bird as large as a hen Pheasant.

Summing up his report of 12 September 1885, Tegetmeier offered the following words of optimism:

I cannot conclude without congratulating Mr. Bateman on the success of the experiment as far as it has yet proceeded. So much harm has been done by indiscriminate and thoughtless acclimatisation, that it is satisfactory to hear that one useful bird has a chance of being introduced under conditions in which other game birds are not likely to do well.

Of course, even if the threat to the tinamous' establishment from shooters could be prevented, there remained the problem of persecution from four-legged predators - most especially the fox, a major hunter of tinamous in their native New World homelands. Yet in his second report, Tegetmeier had dismissed the possibility that foxes would be a danger to them in England:

There is no doubt that an English fox would not object to a bird that is as delicate eating as a land rail [corncrake *Crex crex J)*. The young brood in Brightlingsea are, however, spared that danger, as the M.F.H. of the Essex and Suffolk hounds has, with that courtesy which always distinguishes the true sportsman, granted a dispensation for the season from litters of cubs in the parish.

Tragically, however, Tegetmeier's expectation was not fulfilled. Despite all precautions, the

AHKOHP Alamy Images

foxes triumphed very shortly afterwards, and the tinamous were exterminated. In less than a decade Bateman's hopes for a resident species of tinamou in Britain had been promisingly born, had temporarily flourished, and had been utterly destroyed. By 1896, the entire episode had been relegated to no more than the briefest of mentions in the leading ornithological work of that time. Quoting from *A Dictionary of Birds* by Prof. Alfred Newton and Hans Gadow:

What would have been a successful attempt by Mr. John Bateman to naturalize this species, Rhynchotus rufescens, in England, at Brightlingsea in Essex ... unfortunately failed owing to the destruction of the birds by foxes.

A unique chapter in British aviculture was closed - or was it? In his definitive work *Introduced Birds of the World* (1981), zoologist John L. Long states:

It seems likely that a number of tinamous, other than the Rufous tinamou, may have been introduced into Great Britain, but these attempts appear to be poorly documented.

An event that may have ensued from one such attempt featured a tinamou far from the Brightlingsea area, but sadly the precise identity of that bird is very much a matter for conjecture. On 20th January 1900, *The Field* published the following letter from J. C. Hawkshaw of Hollycombe, Liphook, Hants:

On Dec.23 last, while shooting a covert on this estate, a strange bird got up amongst the pheasants and was shot. On examination it proved to be a Great Tinamu, or, as it is sometimes called, martineta. As Christmas was near, I skinned it myself, with a view of preserving it until I could send it to be set up, and found it to be in excellent condition, with its crop full of Indian corn, which it had evidently picked up in the covert, where the pheasants were regularly fed. The keeper on whose beat it was killed said that he had constantly seen it feeding with the pheasants. If you would be kind enough to insert the above in your columns I hope that I may be able to discover whence this stranger had strayed.

As a footnote to that letter, the editors of *The Field* briefly referred to Bateman's experiment at Brightlingsea, but confessed that they were unaware of any similar trials in Surrey, Sussex, or Hants (Liphook was sited on the confines of those three counties) that might explain the origin of the specimen reported by Hawkshaw.

Not only was this tinamou's origin a mystery, so too was its identity. No description of its appearance was given; the only clues to its species are the two common names, 'great tinamu' and 'martineta', applied to it by Hawkshaw.

Ironically, however, these actually serve only to confuse the matter further, rather than to clarify it.

The problem is that they have been variously applied to at least three completely different species. Both names have been applied to the Rufous tinamou (as in Dr. Richard Lydekker's *The Royal Natural History,* 1894 - 6); but 'great tinamou' is also commonly used in relation to a slightly larger species, *Tinamus major* (native to northwestern and central South America, as well as Central America); and 'martineta' doubles as an alternative name for the elegant tinamou *Eudromia elegans* (inhabiting Chile and southern Argentina).

Was Hawkshaw's bird proof, therefore, of another attempt to introduce the Rufous tinamou into Britain; or was it evidence of a comparable experiment with a different species? Perhaps its existence in the wild was wholly accidental, totally unplanned - simply a lone escapee from some aviary - certainly, tinamous had been maintained in captivity in Britain, with *no* attempt made to release them for naturalisation purposes, by a number of different aviculturists for many years before this event.

Today, even with such established exotica as flocks of ring-necked parakeets (*Psittacula krameri)* flitting through many parts of southeastern England, red-necked wallabies (*Macropus rufogriseus)* hopping across the Peak District moorlands, and golden pheasants (*Chrysolophus pictus)* strutting regally through forest glades in widely dispersed areas of the U.K., it still seems strange to consider that had it not been for an all-too-formidable onslaught by the foxes of Brightlingsea a hundred years ago, Great Britain may well have become home to an entire additional order of birds - that short-legged relatives of rheas and ostriches would have become a common sight by now in the fields and marshlands of England, far removed indeed form their original Neotropical world.

REFERENCES

HAWKSHAW, C. (1900).Tinamu in Hants. *The Field, 95* (20 January): 95.

LONG, 3. L *(1981) Introduced Birds of the World.* David &Charles (Newton Abbot).

LYDEKKER, *R.(Ed.~1894~).The Royal Natural History* (6 vols.). FrederickWarne (London).

MEYER DE SCHAUENSEE, R. (1971)A *Guide to the Birds of South America.* Oliver & Boyd (Edinburgh).

NEWTON[1] A. & GADOW, H.(1894-6) *Dictionary of Birds.* Adam & Charles Black (London).

SICK, H. (198S).Tinannon. In: CAMPBELL, B. & LACK, E. (Eds.), *A Dictionary of Birds.* T. & A. D. Poyse T (Calton). pp.594 - 5.

TEGETMEIER, W. B. (1884). The tinamou as an English game bird. *The Field,* 63 (23 February): 276.

TEGETMEIER, W.B. (1885). The tinamou in England. Ibid., 66 (12 September): 390.

WALKER, C. A. (1985). Tinamous. In: PERRINS. C. M. & MIDDLETON, A. L. A. (Eds.), *The Encyclopedia of Birds.* George Allen & Unwin (London). pp 28 - 9.

Champ: In the Picture

by
Neil Arnold

Before I proceed with this piece concerning a famous monster mystery may I comment that this has only been constructed within the realms of my humble opinion and is not in the view of a scientist a particularly acknowledged cryptozoologist. I am simply an average person with a major interest in all things unknown. Within the four walls of my room, I type with constant frustration, yet hope that someone will scan my words and enjoy them, not discrediting them simply because they have not come from a more experienced or better educated mind.

I am extremely pleased that a forum like this enables ordinary people to express their views and to tell their tales which are just as reliable as anything that emerges from the mouth of any zoologist or any other expert who, in reality, is no nearer the truth just because they have letters after their name. The only thing they have is access to money, which may send them on their expeditions and aid them in their quest for knowledge of this weird kingdom.

The lake - one hundred miles long, a depth of four hundred feet and a width of thirteen miles. Its vein-like rivers end in swampy terrain and all the fresh water within its sheer walls is as black as the night sky. It is a place so vast that it spans two countries; The United States and Canada and is bisected by the national border. From both sides of the border tourists flock to grab their novelty T-Shirts and beady-eyed keyrings. It is also a lake that holds a curse and a legend that has remained strong for many, many years. And what makes it even more exciting is the fact that this place is not Loch Ness but another mysterious lake that appears to be inhabited by large, undiscovered creatures.

Lake Champlain is not a place caught up in the monster trend. Like Loch Ness it has cradled its enigmas for hundreds of years. The merchandise sells like hot cakes, the banks of the lakes are dotted with binocular-holding foreigners and the monster certainly has not failed to keep the mystery alive.

In 1609 Samuel De Champlain fell under the spell of the wonderful lake and was the first person to scan the waters and sight something so graceful, yet monstrous. This sighting would make history and so the flow of sightings would continue and Champlain would almost become a distant relative of Loch Ness.

Casual passers by soon became obsessed reporters and the unknown beast of the depths was soon to be a marked creature with a price tag on its head. But, like every great mystery, it has remained that way and its elusiveness has simply frustrated rather than pushed people to give up in their search. The sightings are way up in the hundreds now and although people aren't sure what type of aquatic creature they are seeing, they are all positive that it is not some known creature.

What I am hoping to do is examine the best two pieces of evidence that concern the legend of Champ.

The two pieces are a photograph taken in the '70s and camcorder footage from the latter part of the '80s. These two pieces are good evidence for and against a large creature. The first picture certainly proves the existence, whilst the second annoys me the more I see it for it is a case of silly misidentification and proves that there are too many idiotic people who will film anything, talk a lot of garbage and discredit the legend even though they probably think they are doing something in the cause for its existence.

Although I despise any overly sceptical view, an opinion has to be honest, so in examining the best two pieces of evidence I will be simply giving my own views although there may be those of you who back every photo. But this cannot be done, just take a look at the Nessie 'surgeon photo' which led people astray for years, although the picture obviously depicted something very stiff.

Champlain is inhabited by some very large fish.

The sturgeon has often been used to explain monster sightings, whilst the Indians believed that the monster in the *chaousarou*, the north American gar-pike. Although some claim that this is the creature that Samuel De Champlain may have seen, most experts believe that he saw a creature with a horse-like head. The *chaousarou* is an aggressive creature that resembles the pike yet has a longer jaw but is far from being a creature with a horse-like head. This beast could well describe the second piece of evidence that I will cover later.

With Nessie, many people have suggested that Loch Ness and other lakes in the northern hemisphere could be inhabited by plesiosaur-like creatures. It does seem that, at least from a number of Champ sightings that this creature is very similar to Nessie.

However, no-one knows how many large, unidentified creatures live in these lakes so giant eels, monster pikes and living dinosaurs are just a few explanations that have been proffered. Some sixty species of fish live in Champlain. Many of them are ganoids, a family of fish-like swimming tanks with their tough armoury and powerful thrashings. However, I cannot cover the species that are beyond our imaginations so let us deal with the possibilities and regurgitate the evidence.

July 5th 1977. Sandra Mansi, her fiancé and two children are paying a visit to Champlain's shoreline. The kids are joyfully wading in the shallows and her fiancé is gathering things from their vehicle when, out of the murky depths Champ rises. A dark body, that leads into a long neck with a horse-like head. The creature holds Sandra in its gaze, its graceful body great in length and unflustered by Sandra's presence it moves slowly. It is then that Sandra's fiancé returns and sees the monster. His immediate reaction is to get the children out of the water.

I. The Mansi photo showing a large,plesiosaur-type beast turning its head.Certainly not a whale or any known species of aquatic creature.

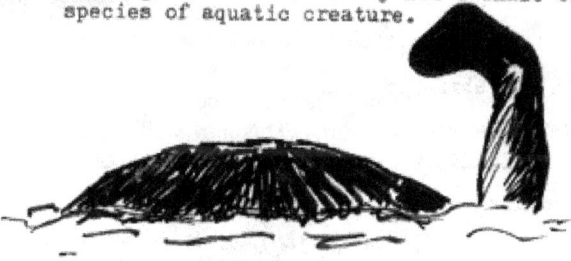

Sandra is hypnotised by the beauty and sheer brilliance of the sighting, and as her fiancé tries to pull her away from the waters edge she takes one excellent photo. The picture captures the creature in all its semi mystical wonder.

After keeping quiet for a while, Sandra then decides to take the photo to an expert who is thrilled by the authenticism and sheer realism.

The clear photo obviously shows a dark creature in full show and about 100 feet away. As always the picture is put under analysis and no-one can find any tampering or way of hoaxing and to this day Sandra is convinced that she saw a monster of the lake. And upon seeing the frame I can only suggest a plesiosaur-type creature and nothing else.

The expert opinion suggests a zeuglodon but that creature, a prehistoric whale, is nothing like this creature. In the diagrams you will see no resemblance, and it makes me wonder what film the experts were looking at. What we have here appears to be a living creature from the age of the dinosaurs that has no doubt spawned from a family that has been undisturbed in the muddy waters.

Yes, the lake has been searched, but its size alone is simply too much of a challenge. It would be absolutely irrelevant to even bring forth explanations such as the sturgeon, for the long-neck dismisses it.

A few people have suggested a whale and that the frame shows a whale turning in the water. Maybe a whale could have entered the freshwater, but so could have a large eel and this is a possibility that I do not mind discussing.

My Loch Ness theory leans towards the eel, but only a truly monstrous eel that defies the laws of science. The creature that resembles a monster pike must also be dismissed as once again the form of the creature suggests a long neck and a fattish body. However, I do believe that the *chaousarou* is valid for some sightings, although the creature is only supposed to grow to a maximum length of ten feet. The monster in the Mansi photo appears to be over twenty feet in length and its overall capacity is greater than most of Nature's creations.

It would seem crazy to suggest that in such an ancient lake only one sort of prehistoric survivor exists or existed. The population may be decreasing, but at one time the waters may have been infested with hundreds of unknown life forms. There may be a number of different species of aquatic, plesiosaur-type creatures that have bred. This may well explain the sightings that have described manes, whiskers and almost feather-like skin, but whatever is down there it certainly has enough food to live on.

This is one of the topics that emerge in the Nessie investigation, but we know by deep scans that there are shoals of fish. However, experts simply dismiss this because they cannot find them, although I'm pretty sure that the plesiosaur is better equipped at finding the food.

Another explanation for Champ is some sort of alligator, and with all the swampy areas this is certainly acceptable. This could even explain the scales and armoured skin as well as the long tail that some witnesses have mentioned. It seems as though the creatures are gathering in the swamps to hibernate and I'm sure that if these areas were not so impenetrable the experts could well discover valuable evidence.

We must remember that the creatures do not have to come to us, they do not have to show themselves and we must not dismiss their existence due to their elusiveness, just look at their home.

And they could well travel on land, but who is to know that they are not frolicking happily within the depths of some vicious swamp.

Another opinion put forward is that the creature is a giant catfish. This is another valid explanation, but when we are trying to explain the Mansi photo it is unlikely. The whiskers are a major feature of the catfish and all over the world the monstrous records are being broken as these fascinating creatures awaken from their dark lairs. A lot of them seem to dwell on the bottom of the lakes, sucking in food and growing bigger every year. Maybe the large heads and boat-like backs are being mistaken for Champ and some people may even mix up the identification of a catfish and a sturgeon so even there, there are problems.

The Mansi photo clearly shows a creature that has become popular in Scotland, in Sweden, in Cornwall and the lakes of Okanagan and Brompton. A few hundred witnesses cannot be wrong. Many experts and sceptics will put forward their more natural explanations but waves, logs, ducks and rocks really are not acceptable with a great deal of sightings. If you go to Champlain and see a slight murmur you cannot scream "monster, monster."

And in the next case it seems as though we have a family intent on seeing the creature and really giving the legend a bad name.

It is July 7th 1988 - almost eleven years to the day of the Mansi incident and the Tappan family are, rather cornily, monster hunting. Apparently one of the family spotted Champ on the 6th so Walter, wife Heidi and daughter Sandy set out on their boat, fully equipped with camcorder, and they only have to wait ten minutes when Champ shows himself. Sandy decides to become hysterical and scream and shouts at her father to film the 'monster'.

And so the film begins.

What we have is a very clear film that scans the water that reflects the pink sky at dusk. Unfortunately, the evidence is very inconclusive and, without sounding sceptical, it seems to show fish mating. All the viewer gets to see are a few small fins splashing about and it is almost typical of a mating ritual. Yet, with Sandy still screaming, Heidi looks around and apparently sees a number of creatures, and believes that where they are situated is a nesting place.

It is then that Walter films something a little better - but not monstrous. About 60ft from the boat, a line of fin-type humps appear. The 'humps' are in a straight line and this seems to suggest that they belong to one creature that is just swimming casually. Walter estimates the length at about twenty feet, although I would say about ten, maybe less.

2. The 45 minute Tappan film that shows a small cluster of of fins within a small area. Certainly not 'Champ'.

In all honesty the film seems to show a pike or maybe a sturgeon with its humps. I would not even consider a monster, and this is a shame considering it is such a good, clear film.

Walter claims that they filmed the creatures for about 45 minutes and that the longer creature was in focus for about 20 seconds. Sandy also claims that she had seven previous sightings and it is now that we delve into the reliability of eyewitness testimony and within this phenomenon we have to trust.

Whilst filming the supposedly 20 foot long creature, Heidi claims that she saw one of the creatures rise from the water and it looked at her.

She says that the creature had a large body and a long neck. However, Walter says that he could not film the creature because by the time he had focused it was gone. Isn't this always the way? It seems that although he could film everything else easily enough, his camera could not film the fully emerged beast, and not even the ripples as it plunged. In my opinion, this family has not filmed a monster and had simply come to the habitat of mating fish, filmed them and then claimed to have seen the real Champ.

I'm sorry that I could not be enthusiastic towards them, but I find it terribly annoying. The two main Champ pieces of evidence are certainly as good as Dinsdale's Nessie film, and the Mansi frame is better than any Nessie photo. But, once again, too many excited people are too eager to submit inconclusive evidence and, surely, they knew full well that they would be ridiculed. Believe me, if I had good footage *I* would submit it despite all the outside pressure, but submitting the film that the Tappan family took was only ever going to attract accusations of an over-active imagination.

Something does seem to exist in Champlain as well as a number of other lakes.

I would say that there is a chance that a good number of undiscovered beasts exist there, but
with the Tappan film I would say that there is no chance that it is a monster, especially Champ. However the Mansi photo is certainly a far more impressive piece of evidence.

All I am saying is use your brain. Sure, film anything that is a little unusual, but we are looking for something far more conclusive than a few lumps. It is this type of identification that fuels the sceptics.

Monster alligator? Living fossil? The mother of all snakes? The daddy of all eels? A prehistoric croc? Whatever it is, it'll outlive us all. And those pesky t-shirts depicting the grinning serpent are pretty accurate because the monster will have the last laugh.

Good.

The Shony, The Lambton Worm, and other dragons of the Northeast

by Mike Hallowell

Whisht lads, haad yer gobs,
An aa'll tell yer all an aaful story.
Whisht, lads, haad yer gobs
An aa'll tell yer aboot the Worm.

On Sunday morning Lambton went
A' fishin' in the Wear.
And catched a fish upon his heuk
He thowt leuked varry queer.

But whatn't kein o' fish it waas
Young Lambton couldn't tell.
He waddn't fash tu carry it hyem
So he hoyed it in a well.

Noo Lambton felt inclined to gan
An fight in foreign wars.
He joined a group of knights that cared
For neither wounds nor scars.

An' off he went to Palestine
Where queer things him befell,
An very seun forgot aboot
The queer Worm in the well.

But the Worm got fat and growed an' growed
An' growed an aaful size.
He'd greet big teeth an' a greet big gob
An' greet big goggly eyes.

An' when at neet he craaled aboot
Ta pick up lots a'nuws,
If he felt dry upon the road
He sucked a dozen coos.

This fearful worm had often fed
On calves an' lambs an' sheep
An swally little bairns alive
When they layd down ta sleep.

An' when he'd eaten aal he could
An' he had had his fill,
He craaled away an' lapped his tail
Curled many times roond the hill.

The nuws of this most aaful Worm
And his queer gannins on
Seun crossed the seas an' got the ears
Of hale and bold Sir John

So hyem he came and catched the beast,
And cut him in two halves,
An' that sewn stopped him eatin' bairns
An' sheep an' lambs an' calves.

The Lambton Worm was written by C. M. Leumane in 1867.

There shall be corals in your beds,
There shall be serpents in your tides,
Till all our sea-faiths die.

Dylan Thomas, *Where Once The Waters Of Your Face.*

The Shony

On September 10, 1999 I completed the manuscript of an article, which I had written for *QUEST* magazine. The article detailed the history of the Shony - a supposedly legendary sea creature that inhabited (and perhaps still inhabits) the North Sea.

The creature sports a thick mane of hair, which stretches all the way down its serpent-like body. The mane is broken by a series of wicked-looking fins, which rise up from its spine.

Mike Hallowell, Hallow'een 2001

The downstairs bar at the *Marsden Grotto* decorated for Hallowe'en.
Opposite: Close-up pictures of the carved pillar

Another carved pillar

By reputation it is the most vicious and powerful sea monster on the planet.

To most the Shony is a mythical beast generated within the fertile soil of Viking mythology. It has, to most folk, no real existence in the objective world. I totally disagree, and I know whereof I speak. The Shony does exist, for I believe I have seen it with my own eyes, and I will never be convinced otherwise. I will detail my experience presently.

The Vikings were convinced that the Shony was real, too. In fact, it was common practice for the captains of Viking longships to "appease" the Shony by slitting the throat of a crewmember and throwing him overboard. To understand why it is necessary to have a brief knowledge of the creature's feeding habits.

The Shony would normally follow ships during a storm, waiting for an unusually large wave to wash an unsuspecting sailor overboard. The sailor would then be seized by the Shony who would take his victim to an underground cave made from the wreck of a sunken ship (some say coral).

There the sailor would wait until suppertime, when the Shony would return and break his fast. Should any of the doomed Viking's friends attempt to rescue him, the Shony would get quite upset and imprison them in place of the first sailor (after having first bitten off their hands and feet), who would then be tossed into the open sea to drown.

In the 12th Century, corpses began to turn up on the beaches of Holy Island with disturbing regularity. These bodies were badly mutilated, and normally had the eyes missing. At first the locals were puzzled, and then reports were made of a huge, nut-brown coloured monster seen swimming nearby.

The monster was very quickly made the culprit, but towards the end of the 13th Century the number of bodies being found diminished, and slowly the local populace began to think that maybe there was no monster after all. Perhaps the mutilations had been caused by the corpses being dashed against the rocks, and the missing eyes could be put down to natural scavengers such as eels and crabs.

But why did the mutilated bodies suddenly begin to turn up? And why did they stop? Surely, if there was a more rational explanation, or some sort of natural process involved, we should still be getting mutilated corpses washed up on Holy Island today.

That idea that there really was a monster lodging off Holy Island is supported by the consistency of the eyewitness testimony. Further, all the sightings occurred around the coastal area of the Island, which faces the mainland.

The first recorded sighting of the Shony in modern times occurred in 1881. One day during that year, a Scottish fishing vessel called *The Bertie* was out in the North Sea during fairly calm conditions. In the afternoon, whilst the crew were busying themselves, their peace was disturbed by a huge marine animal, with a large hump on it's back, which not only reared out of the water but – to the horror of all on board – proceeded to attack the ship.

As the massive animal careered against *The Bertie* time after time, the Captain feared that the vessel would capsize. In desperation he ordered a crewman to fire a rifle at the creature. This stopped the attack for a short while, but then it resumed with even greater ferocity.

For several terrifying hours the crew of *The Bertie* played cat and mouse with the monster. By dusk, the Captain was convinced that it was only a matter of time before the ship sunk beneath the waves. However, just as suddenly as it appeared it returned to the deep, leaving the thankful crew behind to sail home with their story.

In the summer of 1946 dredging work was taking place off the coast of South Shields. A ship named the *Eugenia Chandris* had been wrecked there earlier, and another vessel, a steamer named the *Black Eagle*, was heavily involved in the operation to break her up for scrap.

At some point several crewmembers from the *Black Eagle* spotted a huge "head and neck" rising out of the water to a height of six feet or so. Seizing the moment, the sailors decided to give chase. Fortunately, a motorboat was tied up to the steamer, and within minutes several crew members were making good speed towards the creature.

The leviathan appears to have seen the motorboat coming, and sped through the sea to a safe distance. Then, just as the launch almost caught up with it a second time, the creature sped back again. This cat-and-mouse game continued for quite some time, the Shony always staying tantalisingly ahead of its pursuers, who, according to the account, were trying to see the profile of the creature to compare it with descriptions of *Nessie*. To my knowledge they never did, for the Shony eventually tired of its game and then disappeared under the waves.

My Own Sighting Of The Shony

On August 11, 1998 my wife, my father and I were travelling towards Whitburn along the coast road, near Marsden Rock, when I happened to gaze down at the sea. About thirty yards from the shore was a huge, nut-brown hump just breaking the surface of the water. Although only a small area was above the water level, I could clearly see a much larger area just under the surface. For a second I thought my eyes were playing tricks, and then I shouted, *"Can you see that? What on earth is it?"*

Fortunately, my wife saw it too, although she hadn't the faintest idea what it was. We parked at the next head of cliff and looked down again.

It was still there, but now submerged entirely. Through the waves we could still see its brown colour, although the shape was indistinct. After a minute or so it disappeared. My heart was pounding, for although I had only seen the back of the creature it was no less impressive.

When we arrived home that evening's copy of *The Shields Gazette* was waiting for us on the mat. There, on the front page, was what seemed to be the answer to the mystery; a photograph of a bottle-nose dolphin which had been swimming near the coast and attracting considerable attention. So that was that, then.

But maybe not. The thing which my wife and I saw looked far too big to be a dolphin, and the colour wasn't right for a bottle-nose anyway. And then the telephone rang...

One of South Tyneside's local councillors – who doesn't wish to be named in case his colleagues start to think that he isn't exactly *"knitting with both needles"*, as they say – telephoned me an hour after I'd arrived home about something incredibly boring. After a while the conversation drifted away from business and I happened to mention *Daphne the Dolphin*.

"Funny you should mention that." Said Councillor X. *"I was buying some fish and chips in Ocean Road when I overheard two men in the queue talking about the dolphin. I definitely heard one of them say, 'No way was that a dolphin. What I saw could have swallowed a dolphin in one f-----g gulp.'"*

Perhaps he too saw the Shony.

Worms

But the Shony is not the only "monster" to make its presence felt in the area. The Shony is merely the aquatic version of a similar creature, which is said to inhabit the land near the coast. This creature, known locally as "the Worm", has established for itself a niche deep in the collective psyche of the local people here.

There are many legends and folk-tales about worms [or dragons] in the north-east of England. Folklore expert William Henderson remarked:

"All live yet upon the lips of the people, though of course we cannot presume to guess how long they will maintain their ground against the combined forces of railroads and collieries."

This may seem a curious thing for a learned man to say. Why, one would almost think he believed that the worms and dragons of northern folk-lore *actually existed*. And one would be right, for Henderson and several of his contemporaries most certainly *did not discount that possibility*.

Sir Walter Scott was also a great believer in the existence of giant worms and dragons. In his *Minstrelsey Of The Scottish Border* he suggested that before our lands were cleared of forests and marshes large, serpent-like creatures may have infested our "woods and morasses", and that these animals may well have been responsible for the folk-tales that followed.

Lord Lindsay, in his *Sketches Of Christian Art*, remarked:

"The dragons of early tradition, whether aquatic or terrestrial, are not perhaps wholly to be regarded as fabulous. In the case of the former, the race may supposed to have been perpetuated till the marshes and inland seas left by the Deluge were dried up. Hence, probably the

legends of the Lernaean hydra, &c.

As respects their terrestrial brethren (among whom the serpent, which checked the army of Regulus for three days near the river Bagradus in Numidia, will be remembered), their existence testified as it is by the universal credence of antiquity, is not absolutely incredible. Lines of descent are constantly becoming extinct in animal genealogy."

According to Lord Lindsay, the Worm of north-eastern legend may simply have been a large reptilian creature which has now become extinct. This idea is not as fantastical as it may at first seem.

Theoretically, there is no reason why a large reptile or reptile-like creature could not have survived in the British Isles (although it would have needed a more efficient heat-retaining and temperature-regulating system, biologically, than other reptiles). In fact, there are those who believe that they may not have died out at all.

Even today there are regular sightings of serpent-like creatures in Scottish lochs, as we all know, the most famous being the creature, which supposedly inhabits Loch Ness. What is not so well known is that Ireland has an even larger number of monsters supposedly inhabiting its loughs.

In 1885 a young woman was removing bog-bean from Lough Muck in County Donegal. Suddenly she heard a splashing noise and turned around. She was immediately confronted by the sight of a huge monster with large eyes coursing through the water towards her as if to attack. Not surprisingly she immediately vacated the lough - which is only 1 kilometre long and 785 metres across.

A similar creature was seen in Lough Shanakeever (County Galway) in 1955, not long after an almost identical sighting in nearby Lough Fadda the previous year. Literally dozens of sightings have been reported since then, the last to my knowledge being in 1998. Due to the small size of the loughs it has been speculated that the creatures may actually travel between them by a system of natural underground tunnels and fissures.

It is difficult to believe that all these witnesses are either lying or deluded. However, the cynics and sceptics have a point when they ask, *"Well, if dragons or worms are real, why don't we have more sightings of them, and surely the odd carcass would have presented itself for examination by now?"*

Well, not necessarily.

The existence of the mountain gorilla was only proven last century, previous reports being given no more credence than the reports of yeti and bigfoot sightings are given now; that is, not very much.

One of the arguments put forward against the existence of the gorilla was that no skeletons or carcasses had ever been found. It is also true that in the wilderness areas of the USA the finding of grizzly-bear carcasses is unknown, even when these areas are known to be inhabited by grizzly-bears.

No one denies that gorillas or grizzly-bears exist, of course; so what happens to their carcasses? The carcasses of land-based animals in areas of dense forest do not last very long at all. Predators will strip the remains down to the bones within hours, the bones themselves will be scattered and carried off, if not buried, and within 24 hours there may be little or no sign that the creature had ever lain there after it expired.

The existence of aquatic animals may also be difficult to verify by the finding of carcass remains. In fact, many recently discovered species only came to our attention when they were spotted swimming merrily in the ocean or ended up getting caught in a fisherman's net. The notion that the bodies of dead sea creatures will conveniently make their way to the nearest beach and allow themselves to be washed up on the sands for the next passing scientist to examine is largely a fantasy. Such beach finds *do* happen, but very rarely indeed.

Another factor that can dramatically reduce (or indeed increase) the chances of finding the carcass of an animal is the environment in which that particular species of creature inhabits. The carcass finds of domesticated hamsters is probably around 100%. Hamsters live in cages, and therefore when they die their owners will inevitably find the corpse. But at the other end of the scale we have creatures that are certainly not domesticated and live far from urban areas. Many such creatures have a carcass-find rating of zero.

Now let us return to the supposedly mythological worm. What sort of environment does it inhabit? How does the nature of that environment affect the likelihood of the creature being seen, and what are the chances that we may ever get our hands on a carcass to examine?

The worm of north-eastern mythology appears in a multitude of different guises, physically speaking. However, one universal constant concerns its abode. The worm is *always* described as living in subterranean passages, only rarely showing itself to humans. Nevertheless, the creature's habits were believed to be predictable to a degree. Legends tell of dragons and worms being lured from their lair, or trapped by some brave soul who lay in wait until it emerged to feed. Of course, if the worm did/does live in underground tunnels deep below the surface, then the chances of us finding a corpse to examine are virtually non-existent.

But what of the creature itself? How do early witnesses describe it? A local poet described one creature, which was believed to inhabit the area of "Wantley, near Rotherham" - a village, which I haven't been able to find on any maps, despite numerous efforts - thus:

This dragon had two furious wings
Each one upon each shoulder.
With a sting in his tayl, as long as a flayl,
Which made him bolder and bolder.
He had four long claws, and in his jaws,
Four and forty teeth of iron;
With hide as tough, as any buff,

Which did him round environ.

Some worms are said to have four legs, others two or none. Some possess wings, others do not. Generally, however, they are described as "snake-like" and without many appendages. If they possess any at all they are normally described as "fins" or "wings".

Similar worms are said to inhabit Sockburn (5 miles south of Darlington), and Pollard's Dene (Bishop Auckland). However, the most famous worm of all must be that which was said to have its home at Lambton on the banks of the river Wear.

The Lambton Worm

The Lambton family owned lands and a castle on the banks of the river Wear to the north of Lumley. The Lambtons also had an ancient familial pedigree and were very influential in the locality. Ancient Lambton Castle, which had existed since at least the 13th Century, was demolished in 1797 and replaced by a mansion house.

But it is way back in the 14th Century where our story really begins. One of the Lambton sons - tradition has it that he was called John - was apparently a bit of a wastrel. Not a scoundrel, exactly; just someone who really cared little for spiritual or civil duties, and preferred to spend his time engaged in more leisurely pursuits. John, allegedly, only rarely attended Mass on Sunday mornings. Instead he would go fishing at the nearby river wear, a habit that would catapult him into the history books in the most bizarre way.

One Sunday, John Lambton wandered down through the ashes and poplars, which separated his luxurious home from the river. Choosing a suitable spot, he cast his line into the clear, babbling water and waited. Quite some time later he was still waiting, and so he cursed, withdrew his line and cast it yet again. Still nothing.

Bite, why don't you, damned fishes.

Nearby, the locals and nobles of Lambton were making their way to Brugeford Chapel, and the curses and blasphemies of young John made them wince. Young Lambton could curse like no other when the fish refused to bite.

Eventually a smile came to the young lad's face when his line tightened. Rising to his feet, the young heir started to manoeuvre his catch to the bank. But it was difficult.

This fish must be as big as the Devil himself.

The fish was struggling now, and the water veritably churned. As the youth steadily brought it in closer, several times he saw a part of it flash above the surface as it struggled to escape the hook. There was something odd about this fish, he mused. The brief glimpses it afforded him made him think that it was possibly an eel of some kind, and a large one at that.

A minute later the "fish" was lying struggling on the bank. John approached it, cautiously, his eyes narrowing in disbelief as he studied the creature on the end of his line. This was no "fish" at all.

The creature was approximately three feet in length and olive green in colour. John's earlier appraisal had been right in a sense, for the animal was indeed shaped like an eel, but there the resemblance ended. About six inches from its head the creature sported a pair of appendages similar to the wings of a bat. They were fins of sorts, but like no fins he had ever seen on a fish.

But it was the head itself that revolted John. He had never in all his young life seen anything so hideous. The head was the same colour as the writhing body, and covered in the same scaly, rough skin. Its forehead or brow ridge was very pronounced, and under it sat two large, piercing eyes. They were light green in colour with huge, round pupils. As the beast continued to struggle those eyes seemed to dance crazily in their sockets. Now it was opening its mouth.

You're gasping for breath, you spawn of Satan.

But it wasn't gasping for breath. It was making a desperate effort to bite something; anything, in fact, for this beast had a ferocious temper. Turning its head to John, it displayed a set of obscene, pointed teeth, which looked razor-sharp. The young Lambton also noticed that the creature sported a row of small holes or apertures on each side of its mouth.

By God, if those teeth sunk into your buttocks you'd be aware of it.

So disgusted was John Lambton by the foul-looking creature he was tempted to behead it there and then. Instead, he dragged it across the ground to a deep, natural well which sat nearby. In the creature went, writhing furiously as it plunged to the depths.

A stranger passed by just as John was about to cast his line into the Wear once again.

"And what sport have you had today, Sir?"

"I've caught Old Nick himself if you ask me. Look in yonder well there, and judge for yourself."

The man, who was dressed in a rather peculiar fashion and looked as if he may have belonged to a religious order of some kind, walked over to the well and peered in. Down below he could just make out a writhing, thrashing figure churning up the water.

"I think this is a bad sign, young master; a bad sign indeed."

With that the man disappeared into the trees, leaving a bemused John Lambton to get on with his fishing. After catching two or three he returned to the castle, pausing briefly to look in the well. The creature had disappeared.

Lambton apparently told his family about the "straynge beeste" but then seems to have forgotten about it. Over the next year or two he matured, put aside his youthful follies and excesses and seems to have become a thoroughly respectable young man. The next thing we know of young Lambton is that he went "to fight in foreign parts". Some have suggested that he joined one of the Crusades, but we cannot be sure. (Palestine is almost universally accepted as his destination.) In fact, it is difficult to see how anyone who took part in those bloodthirsty massacres could claim respectability, but that's another story.

During his absence a strange thing happened. Several locals claimed that they had seen a creature of identical appearance to the one caught by young Lambton years earlier. But this one was bigger. *Much* bigger. Brazenly, they said, the beast had curled itself around a huge rock in the middle of the river.

At this point, the factual aspect of the story of the Lambton worm gently and gradually blends with fantasy. Each reader will have to make up his or her own mind which is which.

For some reason the locals seem to have convinced themselves that this animal was the very same creature which Lambton had encountered. This may or may not have been the case, but the assumption that both creatures were one and the same has no logical basis to it as far as I can tell. Whatever the truth, there was now a worm bringing consternation to the good people of Lambton and no one was quite sure what to do about it.

The worm started to make regular appearances. It took to resting at the same spot in the river every day. The castle servants and sundry other folk would gather at the bank to stare. If the temptation arose to throw several stones at the worm then the temptation was resisted.

This thing was big, and the people didn't know exactly what they were dealing with. Who knows, the creature may have magical powers and could bring a curse down on them. Best leave well alone for the time being, they thought. One thing they did notice, and that was that the beast was getting bigger with every passing day.

After a week, one of the locals - perhaps a farmer checking on his sheep - espied the worm resting on the top of a nearby hill. It was there the following evening, too. Now fully aware of the creature's habits, the decision was taken to continue to leave it alone. After all, it hadn't actually harmed anyone.

But then things started to turn a bit nasty. Lambs started going missing. And goats. In fact, rumours started to circulate that a few cows had disappeared under mysterious circumstances too. Fuelled by gossip and speculation, the blame for the disappearances and deaths was laid firmly with the worm. If it wasn't swallowing the cows it was draining their milk. If it wasn't eating the sheep it was taking them home for its young. Why, some were now saying that the worm was so big it had been seen wrapping itself around Penshaw Hill on the other side of the river. Worse, said others. It could wrap itself around Penshaw Hill *three times*.

Within days other tales came from the far side of the river. The worm had "devastated and laid waste" the villages on the north bank. It was causing unbridled havoc. But one had to think of oneself first, and whatever the beast did on the north side was no concern of the people of Lambton. As long as it stayed over there and didn't come over here, they added.

But then the situation changed. A sudden spate of deaths devastated the Lambton family. We do not know what caused them, but they were followed by a rapid departure of staff. Within a short space of time the elderly Lord Lambton was left with only a few family members, a small retinue of soldiers and a handful of servants. His only surviving son, John, was still *"fighting in foreign parts"*.

To make matters worse, the worm now seemed to tire of causing mayhem and destruction on the north side of the river, and now began to turn its attention to the south bank. Specifically, it targeted the estate and property of the Lambtons. Terrified, the aged Lord and his household assembled in council to assess the situation. Much was said during the meeting, but few positive suggestions were put forward until one steward came up with the idea of placing a trough full of milk outside the castle gate (or inside the courtyard, in some versions).

This would, he vouched, act as an appeasement and hopefully prevent the worm from taking more drastic steps to achieve its wont. After all, if appeasing had been good enough for the Shony during the Danelaw, why shouldn't it work with this creature now?

Supposedly the ruse worked a treat, and the monster daily made its way to the trough, drank the milk and departed peacefully. Of course, the castle stewards knew too well that such appeasement would have to continue, and the very next day the worm returned. The trough was already in place, and had, according to records, been *"filled with the milk of nine kye [cows]."*

But there was a problem. The fragile economy of Lambton could not stand such a "milk tax" indefinitely, and so efforts were made to subtly reduce the amount poured into the trough. This upset the worm, who promptly threw a temper tantrum and uprooted a few dozen trees in the local park. The next day the beast received the full "nine kye" worth of milk in the trough.

Eventually some of the Lambton soldiers decided that enough was enough. A succession of them went out to slay the worm, only to end up being dispatched to the Great Beyond themselves. Dispirited and dismayed, Lambton and its inhabitants were fast becoming a community on the verge of extinction.

Of course, in every good yarn there is a hero. In this case it was none other than young John Lambton, the wastrel who came good. Just as the worm was reaching the zenith of its powers, the young noble came home and took charge of the situation.

But Lambton wasn't stupid. If he was to finish off the worm he needed to plan well. He also needed to take advice. He did this by paying a visit to a wise woman or sybil who lived nearby. (The reader will, I am sure, immediately see the similarity between this part of the Lambton story and the legend of the Faeries' Kettle, where the hero also sought advice before taking on the spirits that guarded the chalice. This is not coincidental, as we shall see.)

Lambton's visit to the hag was not altogether a pleasant one. She berated him for quite some time, accusing the young noble of being the cause of their misfortune. If he hadn't dumped the bloody thing down the well all those years ago none of this would have happened.

Nevertheless, she said, what was done was done. She could see that John was stricken with remorse and offered to help.

The crone instructed Lambton to have a blacksmith cover his best suit of armour with spear-heads. This being done, he was then to stand boldly on a rock, which stood in the middle of the Wear. Just as the hero in the legend of the Fairies' Kettle had a "secret weapon" in the form of a bag of feathers, Lambton had one in the shape of a specially designed suit of armour. Apart from this, the noble was armed with nothing but what one scholar called *"providence and his good sword"*.

But there was a condition attached to the sybil's advice. Before taking on the worm, Lambton had to take a vow. Should his mission be successful, he must swear to slay the first living creature he came across on the way home. Should he renege on the vow, the woman assured him that no Lord at Lambton would die peacefully in his bed for nine generations.

Concerned by the vow but determined to go through with it, Lambton made his way to Bruge-ford Chapel. There, he solemnly uttered his oath and said a prayer.

When the blacksmith had finished fixing the razor-sharp spearheads to the outside of Lamb-ton's armour, the young noble donned it and made straight for the river. As soon as he arrived he lost no time in finding a suitable rock to stand upon. It was here where he would have his final showdown with the worm.

Before long, the leviathan arrived and started to glide down the Wear with its head above the surface. As the creature drew within striking distance, Lambton delivered a well-executed blow. His sword caught the worm on the side of its skull and sent it into a paroxysm of rage. Instantaneously it arched its body around the belted knight with incredible speed, enveloping the noble in its coils.

This, of course, was exactly what young Lambton wanted. As the worm squeezed him in its fury, its lifeblood began to drain out of the multitude of small holes now forming in its hide and into the fast-flowing river Wear. All the while, Lambton continued to hack away at the creature with his sword.

One of the legends surrounding the worm - a legend which had sprung into existence even be-fore the young Lord had returned from abroad - was that it could, if cut in two, magically re-unite itself. Lambton knew this, and after a protracted struggle managed to sever its torso into two separate halves and push them both into the river.

Frantically the creature tried to make its two halves connect, but the strong current made them drift even further apart. With its lifeblood still ebbing away the worm had no chance. Within a short space of time it was dead.

Whilst the young knight was battling with the worm, the rest of the Lambton family and their retinue had locked themselves in Brugeford Chapel where they prayed earnestly. Lambton had told his father that, should he be successful, he would immediately blow on his trumpet to alert everyone to his victory. It would also act as a signal for the elderly Lord Lambton to release his son's favourite hound.

There was a good reason for this. The knight had told his father about the vow that he had made to slay the first creature he came across after slaying the worm. To make sure that there would be no tragic accidents, they agreed to release the hound in the sure knowledge that it would make its way straight to its master. Thus, Lambton was assured that the hound would be his sacrifice. Sad, but better that than some unsuspecting villager who just happened to be at the wrong place at the wrong time.

At least that was the plan. Sadly, the elderly Lord Lambton was carried away with the elation of the moment and dashed out to meet his hero son. The first living creature to meet the knight's eyes after his victory was his own father.

The young noble was devastated. What could he do? Slay his own father in cold blood? Never. He would simply ignore it. Once again he blew his trumpet. This time his faithful hound came bounding towards him.

Seconds later it lay dead at his feet, dispatched by the very same sword that had killed the Worm.

But his vow had been broken, and there was an awful consequence to be paid.

We do not know exactly when the events which became written into the legend of the Lambton Worm took place. Sir Cuthbert Sharpe mentions an old manuscript in the possession of the Middleton family of Offerton which states:

"John Lambton, which slewe ye worme, was knight of Rhodes and Lord of Lambton, after ye dethe of fower brothers - 'sans eschew malle' ".

Nine generations counted in the ascent from Henry Lambton MP, takes us back to one Sir John Lambton, who was indeed a Knight of Rhodes. If this Sir John were the same John Lambton who slew the worm, then Henry Lambton would be of the last generation to suffer the curse. Intriguingly, he did not "die in his bed", but passed away in his carriage whilst crossing a bridge which spanned the Wear at Lambton. This occurred on June 26, 1761.

But what of the preceding eight generations before that? Well, Sir William Lambton, a regimental colonel in the service of Charles I, was killed at the battle of Marston Moor. His son was killed in similar circumstances at another battle whilst leading a troop of Dragoons. The other six generations apparently fared no better.

Certain exhibits which apparently verified the entire story of the Lambton Worm were displayed at Lambton Castle. Two stone statues could be seen, for instance, depicting both

Lambton in his specially designed suit of armour and the old crone who had advised him how to defeat the Worm. The trough from which the monster allegedly drank *"the milk of nine kye"* could also be viewed, along with a large piece of the beast's hide, which apparently resembled *"the hide of a bull"*.

So much of the Lambton Worm legend is obviously sheer bunkum. No one seriously believes that the worm was big enough to curl itself three times around Penshaw Hill, although the hill today does have a strange set of circular indentations at its peak which may explain where and how the story began.

Nor, of course, did the creature really possess the magical ability to reunite its torso if it were cut in two. Neither could it have grown to the size described and wreaked the devastation accredited to it.

But we must be careful not to throw the baby out with the bath water here. If we allow for the possibility that the more fantastical aspects of the story were added later, there is nothing impossible about the idea that John Lambton did have an encounter with a real beast, a hitherto unknown species, which only rarely presents itself to humanity.

This beast may be related to the strange and frightening creature that has been seen many, many times in the Irish loughs and occasionally in the Scottish lochs. It may also be related to another creature, which struck terror into the farming community of mid-Wales in the 1980's.

During August, 1988 dozens of sheep and goats in the Welsh hills started to die. At first it was *"one here, one there"*, but then the problem worsened. On one occasion five sheep were found dead at the same time and at the same location. Veterinary experts were totally baffled, as the only clue to the cause of death was a bite-mark on the breastbone of each animal.

The problem began at Bodalog Farm near Rhayader, mid-Wales, which was run by Clifford Pugh and his son Charles. However, within days the creature began attacking neighbouring farms and smallholdings.

In an effort to track the beast locals used hunting dogs, which quickly picked up the scent. Frustratingly, the scent would always lead from the scene of the kill to the nearest river. This led farmers to the opinion that the water was either the creature's natural habitat or, alternatively, that it was using the river as a means of escape.

By October 1988 the killings were still going on, and they began to attract the attention of the national press. Gill Swain, writing in the *Daily Mail* on October 10, mentioned that the beast had been dubbed the Welsh Waterwolf. The name stuck, and the creature now had its own small but significant part in cryptozoological folklore.

One farmer reported that the creature was "teasing" them, and that it seemed to enjoy "leading [them] a merry dance." It is curious that the same thing was said of the Shony in the 1946 incident at South Shields, and also of the creature that harried the fishing vessel *Bertie* in 1898.

Interestingly, the Shony has been linked with another creature from Scotland named the Shaillycot. The Shaillycot, unlike the Shony, is reputed to inhabit freshwater streams, but is also known for its ability to imitate the sound and appearance of a drowning man. When would-be rescuers were just about at the scene, the Shaillycot would disappear under the waves laughing hysterically. Again, it is interesting that both the Shony and the Shaillycot seem to have shared this incredible talent for teasing their observers - a skill which seems to have been inherited by the Welsh Waterwolf.

In August 1999 I spoke to Charles Pugh, and asked him whether the attacks were still going on. "*No*", he told me. "*Thank goodness they've stopped. We never did catch the creature, but we were able to determine that the animals that died had been poisoned by the bite.*"

So perhaps there is some truth to the Lambton Worm story after all, for the similarities between the Worm, the Welsh Waterwolf, the Shaillycot and other dragon/serpent/worm mythological creatures are too great to ignore. They inevitably live in or near waterways, attack goats and sheep and are extremely cunning.

It is also interesting that the Worm, after its capture by the young Lambton, was subsequently dropped down a well or shaft which existed nearby, for there are persistent rumours that a complex of tunnels used to lie under Lambton Castle before it was demolished. It is also rumoured that one of these tunnels actually ran underneath the Wear and then northwards towards Marsden Bay.

As we know, there are also rumours of a tunnel running from Marsden Bay to Cleadon Village, and of a third from Cleadon Village to Tynemouth on the other side of the river Tyne. If these tunnels do indeed exist, then we are faced with the incredible possibility that those with access to those tunnels *could have walked all the way from Lambton Castle to Tynemouth Priory without seeing the light of day.*

But how does all this relate to the central theme of this book? Whoever engineered this vast and complex tunnel system obviously did not intend for it to become public knowledge. A tunnel system known to all is useless. What purpose could it serve? Why not just travel by road?

No; the very existence of such a tunnel complex implies a desire for secrecy of one sort or another; perhaps the wish to travel between A and B unseen and unmolested, or the desire to hide something of great value from the eyes of the world. Of course, having created such a vast network or labyrinth of tunnels one must protect them, and the best way to achieve this is through fear.

The dragon, the worm and the serpent are three potent symbols of dread. Everyone knows that "the Dragon" is really Satan. Does it not say so in the Biblical Book of Revelation? The Serpent? Well, that was Satan too. Did the Serpent not deceive Eve into eating the fruit of the Tree of the Knowledge of Good and Evil? Are we not told this in the Biblical book of Genesis? And the Worm was a satanic symbol too, for the Bible tells us that those who go to Hell will endure such a punishment where *"the fire is not quenched and the worm dieth not."*

That is if you believe the orthodox Christian interpretation of the text, that is, which I don't. But the important thing is that, until recent times, virtually everyone in this country *did* believe it. Dragons, worms and serpents were simply bad news.

The salamander was also a symbol of dread in Medieval times. William Brockie once said, *"We used to be told terrible stories in our childhood, as to what calamities would be sure to happen if such a pestiferous creature was allowed to be bred anywhere."*

To illustrate the dread that this group of creatures could instill we need only look at one report regarding a "monster" which was allegedly brought into the city of Durham in the 16th Century. The account of the incident, which occurred in 1568, is recorded in the St. Nicholas Register:

"A certain Italian brought into the Citie of Durham, the 11th day of June, in the yeare above sayd, a very great, strange and monstrous serpent, in length sixteen feet, in quantitie and dimension greater than a horse; which was taken and killed by speciall pollicie in Aethiopia, within the Turkes dominions. But before it was killed it had devoured (as is credibly thought) more than 1,000 persons and destroyed a whole countrey."

Of course. But despite the fact that the above account has been exaggerated just a tadge, it is obvious that worms, serpents and dragons - not to mention salamanders - would be wonderful deterrents in a culture that had been trained to fear them. It is obvious also that the north-east of England provided just such a cultural milieu.

A Lost World?

TOWARDS A FURTHER UNDERSTANDING OF THE CONCEPT OF HONG KONG AS A LIVING LABORATORY FOR FORTEAN ZOOLOGISTS

by Jonathan Downes

"Study the past if you would divine the future"
Confucius.

PART ONE:
THE MYSTERY ANIMALS OF HONG KONG

EDITOR'S NOTE: Some of the material in this first part has been published elsewhere but it is being re-printed here as an introduction to the ongoing study of the animal mysteries of Hong Kong.

Soon after midnight, local time, in the early hours of the first of July 1997, the Royal Yacht Britannia sailed through the Lyemoon Strait and out into the South China Sea. 156 years of British rule were over, and a new era was dawning for the people of Hong Kong.

Hong Kong is possibly the most enigmatic place on earth. A cultural, economic, political and ethnological cross-roads, it has been described as 'a little piece of the Home Counties tacked onto the edge of Southern China', and it is also, undoubtedly the only part of mainland China to have escaped the ravages of fifty years of Communism.

Despite its reputation as the world centre for *lassez faire* capitalism and 21st Century technology, the twin influences of British and earlier Chinese Imperialism were instrumental in producing the place we know today! Despite its high population density (six million or more people in an area somewhat smaller than the Isle of Wight), and its appalling problems with pollution and urbanisation, it has a rich and varied wildlife. Its position at the 'cross-over' between the tropical and Eurasian geographical areas has given it a unique fauna, which in many ways has a similar relationship with the zoology of the Pacific rim, as Hong Kong itself has with the socio-political and economic infrastructure of the same area.

I spent much of my childhood in Hong Kong, and ever since have made a special study of the animals of the place. They believe that some of what they have discovered has significance far beyond the borders of an otherwise insignificant group of islands in the estuary of the Pearl River, and that the implications have the potential to affect the way we approach the biological sciences as a whole. Cryptozoology is the search for unknown or 'hidden' animals.

The search for bigfoot, 'Nessie' and the abominable snowman is well known, but what is less well known is the search for the answers to many more obscure, but equally interesting zoological mysteries. Since 1988 a startling series of new and exciting animals have been discovered (or rediscovered) in Vietnam - most notably in the province of Vu Quang.

Many cryptozoologists have used these as vindication of the cryptozoological maxim (coined by Bernard Heuvelmans over forty years ago), that "there are lost worlds everywhere". We have discovered that an equally startling series of events has taken place, over the same time scale, in Hong Kong. If the discovery of new species in almost impenetrable rainforest is exciting, the same thing happening (albeit on a smaller scale) in such a heavily urbanised and meticulously explored and mapped place as Hong Kong is by anybody's standards extraordinary! The crab eating mongoose (*Herpestes urva*) was last seen in about 1950, and was even then considered rare. An apparently healthy population was discovered by chance in 1988.

The Chinese otter (*L. lutra chinensis*) was seen for the first time in thirty years in 1990. Two carnivores that had never previously been reported from the territory were also discovered at about the same time. The Javan mongoose (*Herpestes urva*) in 1990, and the yellow throated marten (*Martes flavigula*) in 1994. Two new species of rats (one which may be new to science as a whole), and one new mouse have been discovered since about 1988, and the list of reptiles and amphibians recorded from the territory is growing almost daily.

The peculiar thing about all these discoveries is that they appear to be opposite to all the prevailing trends. The pollution from industrial and human waste has been mentioned already, and several important habitats have either been destroyed or are in imminent danger of destruction.

Habitat destruction has been even more pronounced across the border in Guangdong province, and whereas it is certain that the loss of mainland Chinese habitat has forced more birds into

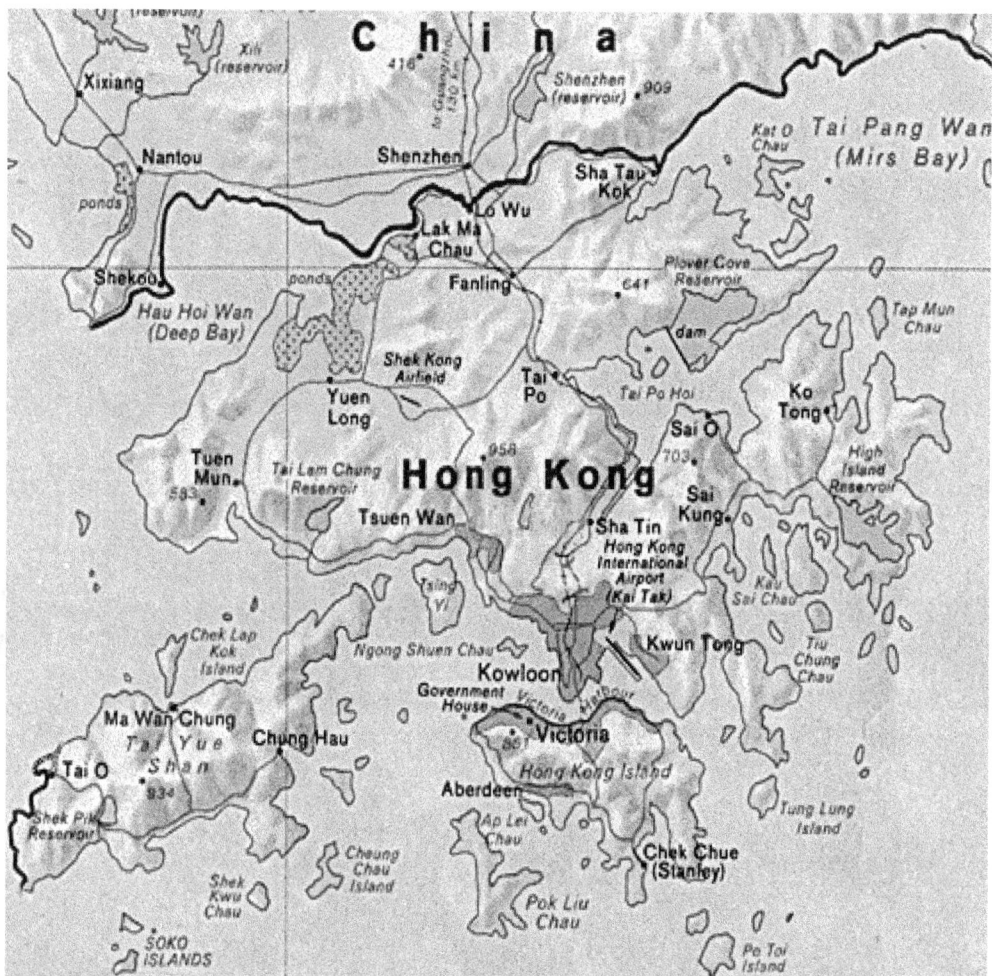

C h i n a

(reservoir)

Xixiang

Xin (reservoir)

416

Shenzhen (reservoir)

909

Nantou

Shenzhen

Lo Wu

Sha Tau Kok

Kat O Chau

Tai Pang Wan (Mirs Bay)

ponds

Lak Ma Chau

Shekoo

ponds

Fanling

641

Plover Cove Reservoir

Tap Mun Chau

Hau Hoi Wan (Deep Bay)

Shek Kong Airfield

Tai Po

dam

Tai Po Hoi

Ko Tong

Yuen Long

Sai O

703

High Island Reservoir

Tuen Mun

583

Tai Lam Chung Reservoir

958

Hong Kong

Sai Kung

Tsuen Wan

Sha Tin

Tsing Yi

Hong Kong International Airport (Kai Tak)

Kau Sai Chau

Chek Lap Kok Island

Ngong Shuen Chau

Kwun Tong

Tiu Chung Chau

Ma Wan Chung

Tai Yue Shan

Chung Hau

Government House

Kowloon

Victoria Harbour

Tai O

934

861

Victoria

Hong Kong Island

Shek Pik Reservoir

Aberdeen

Ap Lei Chau

Tung Lung Island

Cheung Chau Island

Chek Chue (Stanley)

Shek Kwu Chau

Pok Liu Chau

Po Toi Island

SOKO ISLANDS

Coastal Heritiera (Heritiera Littoralis) tree at the seacoast of Lai Chi Wo, Northeast New Territories Islands, Hong Kong

certain Hong Kong habitats than were seen previous to the urbanisation of rural China, it appears unlikely that the same ecological restrictions could have forced a tide of refugee mammals across the border into Hong Kong.

As recently as the 1960s, the border with the rest of China was a barbed wire fence with identical swathes of paddy fields on each side. Animals could quite easily cross from Chinese to British territory, and on a number of occasions they certainly did. For over twenty years, however, they have been replaced by a grotesque and essentially lifeless concrete jungle. The industrial estates of the Shenzen Special Economic Zone provide an effective barrier to any mammal population expansion.

Could man have introduced these animals? Although all the species cited are known as food animals, the specimens recorded have been in perfect condition, suggesting that they had bred in the wild, and in Hong Kong itself. They showed no signs of having been trapped or caged. (Creatures for sale in the wild animal food markets often suffer from horrific injuries). None of these species are commonly kept as pets so the hypothesis used to explain the existence of exotic animals in so many other parts of the world can also be discounted.

The question remains however. Where have these animals been all this time? The zoology of Hong Kong has been meticulously mapped for over a century and it is inconceivable that entire species could have eluded so many generations of keen amateur and professional naturalists, who for a century and a half kept meticulous records. When one examines this wealth of source material, one finds that these animal anomalies are only the latest in a long line of strange and mysterious creatures which have turned up in Hong Kong to leave a few tantalising clues to their existence before disappearing again into the limbo of zoological obscurity.

It is generally believed, for example, that there are only two species of Asiatic giant salamander, one in China and the other in the mountain streams of Japan. In 1922, the type specimen of what was described as a third species - Sligo's salamander (*Megalobatrachius sligoii*).

It was discovered in a burst drainpipe on Hong Kong Island itself, and it had a smoother skin and a flatter head than the mainland Chinese giant salamander (which has never been recorded from Hong Kong anyway) and according to some reports reached up to five feet in length. Other specimens were discovered, both in Hong Kong and elsewhere in China, and they were exhibited live at London Zoo.

In December 1941 Hong Kong changed forever with the advent of the invading Japanese, and many records were irrevocably lost. Much habitat was also destroyed including the mountain streams where, Jon Downes believes, this remarkable creature would have lived.

After the war, the political upheavals in the Far East precluded much serious zoological investigation, and despite the fact that the type specimen still exists, preserved in spirit at the British Museum (Natural History) in London, the species has been quietly forgotten. It is as if it never actually existed. There are several other problematical animals, which are equally as interesting as Sligo's salamander on record.

These include a giant tortoise, at least one species of terrapin, and an animal described as a 'six legged snake'. There is at least one report of a sea-serpent seen by Chinese students in about 1968 and other less concrete sightings abound. Richard Muirhead has discovered records of an Indopacific crocodile (*Crocodylus porosus*) in 1912 and 'wolves' and badgers and a 'black panther' from the 1920s. Needless to say there have been no other records of these species recorded since.

The conclusion that we have reached is that Hong Kong is an invaluable 'living laboratory' of equal importance to those in the more zoologically recognised areas such as Vu Quang and the Galapagos Islands. This is because, although it is generally accepted that the fauna of a specific area is never static, the fauna of Hong Kong changes at a far faster rate than that of most other places due to the high degree of biological, geographical and socio-political dynamism in existence in this former British Colony.

What is more important, however, is that Hong Kong provides an object lesson to zoologists who are convinced that there are no more significant zoological discoveries to be made. If the archives of such a tiny British Colony have yielded so many surprises just imagine how many discoveries could be made if the archives of, say, British India - an infinitesimally larger and more diverse habitat - are examined properly! And, if the tiny remnant forests of Hong Kong can even now reveal surprises, imagine what is lurking undetected in the depths of Amazonia or Equatorial Africa.

It has been suggested that because of the peculiar circumstances of the appearance of some of the 'new' species discovered in recent years, that accepted models of population and species dynamics cannot be responsible. Quasi fortean theories of species translocation and even animal teleportation have been suggested as explanations, but we don't go that far. Hong Kong is a strange enough place already without having to write those sorts of ideas into the equation!

THE MYTHICAL MERMAID?

A dugong was harpooned by local fishermen in 1941 and its carcass was brought in triumph back to Hong Kong. These magnificently primitive marine mammals are not supposed to exist in the South China Sea, but at least one specimen obviously did. It has been speculated that sightings of these grotesquely human-looking creatures are responsible for the reports of mermaids, which have been made for centuries by fishermen in the South China Sea. Only a few years ago a Hong Kong fisherman reported by radio having caught a 'mermaid' in his nets. A vast crowd turned up on the quay waiting for his return, but no-one turned up. The rumours quickly went around that the fisherman's boat had capsized under mysterious circumstances. It is tempting to suggest that this could be circumstantial evidence for the Chinese superstition that catching a mermaid is bad luck, but it is more likely that it was a hoax.

MONITORING THE SITUATION

The largest lizard known in southern China is the water monitor *(Varanus salvator)* which grows up to a length of between two and three metres. There are a few records from the paddy fields of rural Hong Kong but it is generally agreed that these are vagrant specimens that have wandered over from the mainland.

On the 21st January 1930 a lady walking along Lugard Road on Victoria Peak – the largest mountain on Hong Kong Island - was frightened when she saw what she thought was a "miniature crocodile". With the help of a passing policeman and some Chinese coolies, and a "Japanese gentleman who was passing" they cornered the creature. With great presence of mind the un-named Japanese gentleman took off his coat and threw it over the animal. The lizard later allowed itself to be dumped in a sack and to be taken to a Police Station, and ultimately to the Botanic Gardens where it 'was placed in a cage`.

It was examined by Dr Geoffrey Herklots, the most famous naturalist then living in Hong Kong. His description was:

Total length - 22 feet 10¼ inches, head: 6 inches, tail: 1 foot 6¼ inches.

Breadth - At neck 2¼ inches, middle of body 6 inches, in front of hind limbs 2½ inches, middle of tail 1 inch.

Depth - Base of tail 2 inches, groove along back and beginning of tail, ridge along rest of tail.

Colour - Above brown-grey, or deep olive, with yellow spots or hands, below a dirty yellow,

Herklots noted that this was only one of several records of strange lizards seen both on Hong Kong Island and on the mainland at the time. As a child in Hong Kong I saw a similar creature on the Island during 1968. Various suggestions have been made for the identity of these creatures. It was initially identified as *Varanus bengalensis*, a species that isn't actually found in China. It was also tentatively identified as an African species – *Varanus albiguaris*. The surviving photographs, however, suggest that it was not either of these species. It is also certain that it is not the indigenous *Varanus salvator* so what was it?

Although these days, exotic animals from all over the world are kept as pets, and escapees undoubtedly can and do become established in the wild, the international trade in exotic reptiles was almost non-existent seventy years ago. Therefore the suggestion that the lizard, which died soon after capture, was an escaped African *Varanus bengalensis* can, I think, be discounted.

BIG CATS

The most exciting animal ever to have been recorded from Hong Kong is undoubtedly the South Chinese tiger *(Panthera tigris amoyensis),* and even the records of this unmistakable animal are shrouded in mystery. The last 'definite' tiger record was in 1947 when one was seen, by the Bishop of Hong Kong, walking across his garden. Even a Bishop's evidence, however, is open to doubt. He described the pug marks left by the animal as being six inches

Photograph of skin of Leopard, *Felis pardus.*

Plate 11.
Printed by S.C.S Phot.

The tiger was not the only big cat recorded from Hong Kong; this leopard was shot in 1931, and another was shot in 1959

Is *Macacus sancti-johannes* an extinct race of the rhesus macacque?

across which is far larger than the pug marks ordinarily left by a South China tiger. So, if it wasn't a tiger, what on earth was it?

Another tiger was shot on Hong Kong Island during 1941, by guards at the P.O.W camp at Stanley. This record is also highly mysterious, because the Japanese military authorities made a propaganda exercise out of the whole affair, utilising an ancient Cantonese belief that the advent of a tiger meant political change for the better. It has been suggested that this unfortunate animal was not a *bona fide* wild tiger at all, but a tame animal released by the Japanese merely so they could shoot it and gain political 'brownie points'.

TOO MUCH MONKEY BUSINESS

Another mystery surrounds the exact status of the monkeys that live in various parts of the territory. In the mid 19th Century Robert Swinhoe described them as *Macacus sancti-johannes,* a species endemic to the colony. These animals had distinctive 'pig' tails, but had apparently disappeared by the turn of the century. Monkeys of other species started to appear soon after the First World War, and at the last count there were five different species, living and interbreeding in the wilder parts of Hong Kong. Some of them are descendants of animals liberated from private collections. The provenance of others is less certain. What we do know is that none of them have tails like pigs. Did *Macacus sancti-johannes* ever actually exist? If so, what happened to it? The mysterious disappearance and re-appearance of the monkeys of Hong Kong are just another zoological enigma of this strange and wonderful land.

PART TWO:
A ZOOGEOGRAPHICAL PERSPECTIVE

Over the years I have been asked repeatedly, how there could be any 'mystery animals' in a small place the size of Hong-Kong? After all, the island of Hong-Kong, ceded by the British from the Chinese in 1843, is only thirty square miles in area.

The tiny portion of the mainland known as Kowloon, which was ceded eighteen years later is only three and a half square miles in size, and even the disputed lands leased from China on a ninety nine year lease in 1898 is only an area of 355 square miles. The border with The Republic of China is only seventeen miles long and most of the two hundred or so (no-one seems to know the exact figures) islands that make up the colony are barren and rocky.

The answer is that nearly all of Hong Kong's animals are a mystery. The fauna is rich and varied, and because of the changing political, and socio-economic roles of the colony, because of its place on the borders of Paelearctic and Tropical vivisystems, and possibly most important, because of its unique position as a 'cross-roads' at the mouth of the Pearl River its zoofauna is constantly changing.

Its name and status have also changed over the years.

During the late spring of 1968 I was nine years old and sitting in a geography lesson at Peak School on Hong Kong Island. I was not a good pupil, and my schooldays were far from successful, but for some reason this particular geography lesson remains fresh in his memory thirty-three years later.

My teacher, a Mrs Alexander, a lady of whom I can remember almost nothing apart from the fact that we lived in the same block of flats at Mount Austin Mansions, was explaining about the geography of Hong Kong. It was only then, after having lived in Hong Kong for the whole of my conscious life that I realised quite how small it was, and began to be aware of the peculiar nature of the land in which I lived.

"By the time you are my age", said Mrs Alexander, *"Hong Kong as you know it will not exist any longer"*. This was an extraordinary concept for a nine year old to grasp, and so I missed the next sentence while I grappled with it.

"Hong Kong is not a country in itself", she continued, *"it is run by people in England, another island on the other side of the world"*.

I knew this of course, my father worked in that unfathomable entity of the 'grown-up' world called 'The Government', but it was only then, as Mrs Alexander pointed out the relative positions of Hong Kong and the United Kingdom on the map of the world, large portions of which, even then were still coloured pink, that I began to grasp the distances involved. As the bell rang for the mid-morning break Mrs Alexander finished her lesson with words that remain fresh in my mind nearly three decades later.

"Hong Kong is not a country, it is a Crown Colony. When you have boys and girls of your own there won't be any colonies left, and you will be able to tell your children how you lived in a little piece of history".

For political reasons although Hong Kong was still legally a Crown Colony, it began to be known as a 'Territory' somewhere in the middle of the decade after Mrs Alexander's lesson. I was not aware of this politically correct nomenclature. Decades later I managed to mortally offend one correspondent by referring to 'The Colony of Hong Kong' in a letter about sea serpents.

It is, of course, now a Special Administrative Region of the Peoples Republic of China, but for reasons of etiquette, the authors use the term 'region' except when quoting directly from another source, or referring to Hong Kong's position in Colonial History. The name doesn't matter because the place will remain. Despite the encroachments of Communism, urbanisation and pollution, the bamboo snakes will still live in the thickets and people walking in the hills will still see the scurry of paws as ferret badgers disappear into their holes in the ground. The animals of Hong Kong, who care little for politics, will remain as enigmatic as ever.

Over the last century and a half, in fact since before the British arrived, people have been writing about the wildlife of this tiny, and geographically insignificant countryside. The interesting thing is that each account is significantly different from the one before, and that practically everything written about the wildlife of the colony, by outsiders at least is wrong. For example, Hughes included this passage in his 1968 book, *Hong Kong - Borrowed place, borrowed time*:

"There is enough wild-life in the colony's unsettled and woodened heights - to say nothing of wretched mongrels running loose, and puppies being fattened for a Cantonese banquet delicacu - to keep a very British Society for the Prevention of Cruelty to Animals alert, active and anxious. There are barking deer, and scaly ant-eaters, and ferret badgers and civets and porcupines, eight species of rats and mice, two of shrews and at least ten species of bats, including the dog-faced fruit bat, which for some doubtless good reason insists on nesting only under Chinese fanpalms. The wild-boar has been killed off, like the crab-eating mongoose, the rhesus monkey, and the Communist's faithful friend the Dhole - a red, and presumably not a running, dog.

There are also enough birds to keep a very British Society of Bird Watchers expectant, binoculared and exultant. There are grebes and herons, shrikes and drongos, babblers and kingfishers, hawks and eagles, crakes and rails, owls and cockatoos, (also a francolin whose sharp, prolonged call, is said by the imaginative to sound like, `Come to the Peak Ha! Ha!').

But there are no seagulls in the harbour: Hong Kong throws away nothing that can be eaten or sold!"

Richard 'Dickie' Hughes was a journalist with impeccable credentials (Wilson p.86). His greatest 'scoop' was the unmasking of Burgess and Maclean in Moscow, and his writings are always perfectly researched. For a man like him to fall down on such basic facts seems extraordinary! There are several mistakes, just in this short passage. Even in 1968 there were at least seventeen species of gull and tern known from within the borders of the territory, (Herklots noted, the same year that *"For the most part gulls are winter visitors to Hong Kong harbour",*) the crab-eating mongoose was rediscovered in the late 1980`s and the wild boar, even in 1968 was not entirely extinct. Since then its numbers have recovered dramatically.

It is hard not to draw the inference that the wild life is so little known, and of so little importance even to some of those who claim to be interested in the subject, that it has often been relegated to a position where it is used merely as an allegory to support the writer's own viewpoint on the territory and its history and/or economics and/or politics, and the facts about its zoology have been relegated to a humble second or even third place.

The truth is, that all Hong Kong's animals are a mystery. There are glaring gaps in our knowledge of even well known species, there are enough zoological anomalies to keep even the most jaded investigative zoologist on his toes, and there are even a few real 'cryptids' or true unknown animals left to investigate.

For those interested in the less tangible aspects of fortean zoology there is a fair smattering of what Loren Coleman and Mark Chorvinsky have called 'high strangeness', to season our zoological broth with some exotic and disturbing flavours.

After all, Hong Kong is possibly the most fortean place on earth!

Hughes was, however right about the Chinese francolin *(Francolinus pintadeanus)* a bird similar to the partridge. It is perhaps the one creature in the zoofauna of the colony that is mentioned, usually accurately in every account of its wildlife.

In 1933 Robert Simpson wrote a poem about this singular bird.

Come to the Peak Ha! Ha!

Some hyperaesthetics scream out to suppress,
cracker firing while others direct their rebukes,
At Aeronaut zoomers, some try to redress
The false note of gramophones, saxophones, ukes.
My bete noire is that oiseau de mauvais augure,
who addresses in accents impeccably pure,
`Come to the Peak Ha! Ha!

`If my sunday is blank in my mid level flat,
For an afternoon`s nap, I sometimes extend.
I toss and I turn, try this side or that
,but repose is a mood that I can`t comprehend.
`Oh is there a breeze in this island?` I cry,
and that fiend of a bird makes the scoffing reply;
`Come to the Peak Ha! Ha!

`And when I`m invited to tennis up there,
and climb on a tram-seat at Kennedy Road,
should it chance that a dripping wet fog fills the air,
and the universe feels like a vault for a toad.
As I take my last mouthful of dry atmosphere,
the bird on the hillside lets out a grim jeer
-`Come to the Peak Ha! Ha!

`I play golf at Fanling on a dazzling day,
when the ball is a basilisk`s eye in the grass,
if I`d kept my eye on`t, I venture to say,
t`would have blasted my eyesight,I was out of my class.
I foozled my putts; sclaffed shots off the tee,
and each time I swore the bird shouted with glee
`Come to the Peak Ha! Ha!

`In a thunderstorm once on Dairy Farm Hill,
I sailed with my car, even backed in reverse,
Get out and get soaked? Or wait here until

the tropical storm clouds of summer disperse?
That was the question I asked of the night,
that bird in the coarse grass replied in delight,
`Come to the Peak Ha! Ha!

`From Stanley or Shek-O, to Castle Peak Bay,
Cheung Chau to High West or the Tai Mo Shan crags,
be you outdoor or indoor you can`t get away,
from this bird which incessantly, tirelessly brags.
Who first was the wretch with the disarranged brain,
that instructed a bird to proclaim the refrain
`Come to the Peak Ha! Ha!

According to Herklots, also writing in 1968:

"The call of this bird is the most familiar of all bird calls in Hong Kong. Who does not know the cock`s loud challenging crow `Come to the Peak Ha! Ha!`, but far fewer know the bird by sight because it is shy and retiring except in the spring. When the first fogs of spring cloth The Peak with a cold, grey mantle, in the early morning the cock bird seeks a commanding rock or stump of tree and utters his challenge, another cock answers and soon several may be heard calling from several directions".

This is perhaps the only animal that you will meet within the pages of this book, which is not shrouded in mystery, for much to the surprise of the present authors (and indeed everyone who has been involved with this book over the four years it has taken to write), Hong Kong and its wildlife are very mysterious indeed.

Perhaps the biggest mystery about Hong Kong and its animals is why in a zoofauna which contains such diverse and exotic animals as pangolins and tigers, and a cryptofauna which ranges from dragons to sea serpents, why an insignificant bird superficially similar to a partridge is, perhaps the best known animal in the territory!

Hong Kong, at least within recent historical times, has often been seen as a cultural backwater of outdated colonial shibboleths. This could not be further from the truth. It has always been a microcosm of global culture, reflecting in its very diversity, the trends currently operating throughout the world. From a zoological point of view, as we approach the third millennium, with the watchwords of bio-diversity and an aim of self-sustaining vivisystems, Hong Kong, a crossroads of North and South, a little pocket of the West in the Far-East has never been more important. When discussing the more fortean aspects of our subject it is perhaps important to remember that Hong Kong is merely a peninsular of the Chinese province of Kwantung (now GUANDONG), and that Kwantung has always been a particularly peculiar place. As Frank Welsh wrote in 1993: (p.13)

"Kwantung is an odd region, regarded by the rest of China with suspicion and disdain. The Sung Emperors, who ruled China from 960 - 1279, were worried by the great number of Wizards and sorcerers in the city, and issued special edicts forbidding human sacrifices to be made to demons".

Welsh continues, attributing this racial distrust and feeling of cultural alienation to Kwantung's remoteness from the rest of China. Canton is after all two thousand kilometres or so from Beijing, and is separated from the older capitals of China by mountain ranges and impassable country. He also propounds another theory to explain, and one that will become very important at various times in this book.

"It seems that Cantonese will eat anything - and it is not all that easy to find food unacceptable to Chinese cookery - bats, tortoises, raw monkey brains and new-born rats for example" (p.13)

A amusing, possibly racist, and certainly fortean joke along the same lines was recounted by Lee & Brown (1993):

"The story of the day that an alien lands on earth and appears before an Indian and a Cantonese person - the Indian prostrates himself before the alien and prays to it, while the Cantonese person hurries off to consult a recipe book". (p.364)

A 1990 account by Dick Wilson, reiterates Welsh's claim that the indigenous inhabitants of Hong Kong will eat anything, and also provides evidence to suggest that gustatory demands are directly responsible for the increasing rarity of some of the region's animal life: (p. 79-80):

"Annie Wu catered for gourmets in the 1980's and considered a menu of bear paws and pine seeds, braised elk trunk with boiled turtle, snow frog fat and moose nose to be a good starting point. (....) When the Mandarin Hotel celebrated its twenty-first birthday, it gave a dinner which stretched over three consecutive evenings, and included wild duck's tongues, snow

frog's ovaries with ginseng, sturgeon maw, civet, sea cucumber, crane and cordyceps (the lat-
ter being a worm in winter and a plant in summer).

Other delicacies that occasionally figure in formal banquets for serious students of gastron-
omy are elephant trunk, stork, seal, deer tail and sturgeon bowels. Government inspectors in-
vestigate restaurant kitchens regularly, and in one year recently they found 12 bear paws, 16
scaly anteaters, 25 giant salamanders and 117 birds of prey waiting to be cooked and served".

Asiatic bears, pangolins and giant salamanders are all endangered species, as are many birds
of prey. These are all animals that we will meet again during this book. The authors are stress-
ing these seemingly bizarre exercises in gastronomy because as will be seen they are integral
to the changing fauna of the region. The Chinese abhor meat and fish that is not freshly killed
and therefore most food animals are alive when sold at markets.

Goodyer wrote in 1992:

"A recent visit to the Guangzhou food market found the following mammals for sale: sixty five
hedgehogs, twenty five bamboo rats, forty six flying squirrels, thirty five Racoon Dogs, one
European Badger, seventeen Hog Badgers, forty-nine Ferret badgers, one hundred and forty-
two Masked Palm Civets, three Small Indian Civets, four Crab-eating Mongooses, one Tem-
minck's Cat, four Leopard Cats, thirteen Wild Boar, four Indian Muntjacks, six Chinese Munt-
jacks and two Water Deer. At the present rate of trapping in China, it will not be long before
much of China's remaining mammalian fauna is wiped out".

Dick Wilson's evidence suggests that the gastronomes of the territory may be partly responsible for the decline in numbers of the Chinese pangolin, *(Manis pentadactyla)* a rare and curious animal endemic to the territory but are undoubtedly also responsible for the arrival of several 'new' species.

In 1992, Goodyer provided a spark of hope when he wrote:

"With the 'westernisation' of Hong Kong Chinese youth, the desire to eat wildlife may be de-
creasing, although there is a worrying tendency among the rich to display their wealth by
holding 'wildlife banquets'".

Hong Kong is perfect ground for a fortean zoologist. It is part of the strangest region of China,
it has practising devotees of religions largely vanished from the rest of the country, each with
their own peculiar pantheon of semi corporate gods and daemons, and it has a population with
such bizarre tastes in food that exotic and unusual creatures are imported regularly for food,
and equally regularly escape into the countryside, introducing new and varied spice to an already exotic zoology.

Although this book is essentially one of zoology, we are taking a fortean approach to it. Many
of the animals described herein are well known either from Chinese folklore or from quasi-
religious iconography. We feel that although, as Wilson (1990) wrote organised religion itself
is relatively unimportant in Hong Kong and involves only a relatively small section of the

population, nearly everyone is intensely superstitious, and as Wilson said the country is immensely 'respectful of the occult'. (P.58). Feng Shui and numerology are common aids to business as are various forms of soothsaying, and the spirits of the ancestors are omnipresent. We make no apologies, therefore, for covering aspects of Feng Shui and Chinese horoscopes, which involve animals at various parts of this book.

The animals of Chinese mythology are rich and varied, and have to be discussed alongside their flesh and blood counterparts. One has to remember that Imperial China was, in its way as multi-cultural as modern Hong Kong, and as Welsh noted, (p.17) the Emperor Ch'ien Lung brought Chinese Turkestan (now SINKIANG) into the empire and forced 'The Gurkhas of Nepal, the Nepalese and most of Indo-China to acknowledge the Emperor as overlord' such animals as elephants (which have not inhabited China since prehistoric times), and lions (which never lived there) are familiar items in the Chinese bestiary.

When the British first arrived in 1841, Hong Kong Island itself was practically uninhabited, apart from a small fishing community, although there was a well established population in what was later to become the New Territories. The first human inhabitants of the area seem to have arrived in the first millennium B.C, and it is interesting to note that even at this early date, what artefacts have been found have a distinctly 'Chinese' look, although these early inhabitants of the region were not of the Han race.

It is unclear when the coastal regions of southern China first entered Imperial rule, although there are few records before the eleventh century A.D. The Han had, of course, colonised the area long before this date, but the first records to be found date from the Sung dynasty.

The whole region had been completely depopulated in 1662. According to a contemporary report from the East India Company:

"...nor is there any certainty of trade in any part of China under the Tartar, who is an enemy to trade and hath depopulated all the vast quantityes (sic) of islands on the coast of all maritime parts of Chyna (sic) and 8 leagues from the sea, merely not to have trade with any". (Morse)

Ironically even this episode of Chinese history is shrouded in mystery as Brown and Lee claim that:

"The Tsuen Wan area was completely depopulated following the orders of the seventeenth century Manchu government to abandon the coastal villages in response to constant pirate attacks". (p.145)

In an interesting parallel to the shifting fauna of the region that is discussed throughout this book, the human population changes its habitat, in what is perhaps a more spectacularly obvious way than the authors have encountered anywhere else in the world. The depopulations of the coastal regions during the seventeenth century that we discussed above are one obvious example, but Brown and Lee (p.187) provide a more recent parallel:

"Some of the smaller islands around Lantau provide ample evidence of the rapid depopulation of the outlying islands, as people move to Hong Kong Island and the New territories in search of work.

the most extreme example of depopulation is the little archipelago of the Soko islands, south of Lantau which once supported viable communities linked to Hong Kong by ferry. As the inhabitants moved out, the islands became the target of weekenders on private yachts seeking an isolated anchorage. In 1989, the last two residents, an elderly couple living on Tai A Chau, the largest island moved out, renting their home to the government who turned the Sokos into a Vietnamese `reception` centre - a desperate move to alleviate the accommodation shortage for that summer's newly arrived Vietnamese. there was no shelter or fresh water on the Sokos at the time, and four thousand people were crammed onto an overgrown, rubbish laden Tai A Chau - a sad way of repopulating the islands".

Wilson (1990) chronicled other mass population changes:

"In the 1840`s the entire Chinese Community left in a body for the mainland, in protest against the Governors passing a law to register them and charge a fee for it. In the 1890s, half the population left Hong Kong because of an outbreak of Bubonic Plague. There was an exodus in the First World War and another during the Great Depression in the 1930`s. The most recent recurrence came in 1941 when the Japanese Army conquered Hong Kong and half its people thought it prudent to pack its bags and go to ground in the interior of China". (p56)

It is this social and sociological mobility amongst people and animals, and the aptitude for adaptation that is shown by all that makes Hong Kong such a socially and biologically diverse environment. These mass movements of people means that environments suitable for animal habitats are ever changing.

What was once a flourishing community can be a wilderness only a few years later, and the influx of a new race of people to the region, like the Hakka, the 'Jews of China', migrant people who originated in Shandong, but left in the third century BC (Wilson p.20) who arrived in the New Territories about three hundred years ago, and the most recent influx of Vietnamese immigrants during the 1970`s each have the potential both to change the environment and bring new animal inhabitants into it.

Perhaps it is not coincidental that two species of Indochinese squirrels (*Callosciurus flavimanus thai* and *C. pygerythrus styani*) now well established residents of the territory first appeared at the same time as the Vietnamese 'Boat People' from the same geographical region.

The present authors have uncovered dozens of accounts of this phenomenon of population shift, but have manfully resisted the temptation of listing them all in full. Even without resorting to the contemporary mathematical model known as `The Butterfly Effect` it seems undeniable that these mass movements of people, must have a direct effect upon their environment and therefore upon the other animals that live there.

Cryptozoology is, according to its 'father' Dr Bernard Heuvelmans, the 'study of hidden animals'. Such animals are defined as species hitherto unknown to science. In the years since Heuvelmans first propounded his new science it has been expanded by various practitioners to include the study of animals thought to be extinct, and also the study of animals of known species found in places where they are not hitherto known to exist.

Because of its very nature the search for 'hidden' animals is full of surprises. That is probably the main reason that people like the authors do what they do. 'Searching for hidden animals' doesn't earn a great deal of money, nor does it engender a great deal of public respect, but it keeps its practitioners on their toes.

Both the authors have been familiar with the mammals of Hong Kong since early childhood, but to their immense surprise and delight their initial research for this book turned up two new carnivore species from the region only discovered since 1985.

The authors then compared the lists of mammals known from the colony from different sources, and extrapolated some surprising conclusions.

The earliest record of Hong Kong mammals that we have found is from 1893:

"The zoology of Hongkong (sic) is limited, as regards wild animals, to a small deer, a badger and a species of wild cat, but these are not numerous..."

Ironically one has to go forward over half a century before one can find an attempt at a complete listing. Herklots (1951) listed the following mammal species:

INSECTIVORES

Large musk shrew or house shrew (*Suncus murinus*) (Linn).

PRIMATES

Rhesus monkey *(Macaca mulutta)* (Zimmermann)

CARNIVORES

Mustelids

Common ferret badger (*Helictis moschata moschata)* (Gray)
Eastern Chinese otter (*Lutra lutra chinensis)* (Gray)

Canids

Wild red dog (*Cuon javanicus lepturus*) (Heude)
South China red fox (*Vulpes vulpes hole*) (Swinhoe)

Viverrids

Rasse or little spotted civet (*Viverricula malaccensis malaccenis*) (Gmelin)
Chinese civet (*Viverra zibetha ashtoni*) (Swinhhoe)
Masked palm civet (*Paguma larvata larvata*) (Hamilton Smith)
Crab eating mongoose (*Herpestes urva*) (Hodgson)

Felids

South China tiger (*Felis tigris amoyensis*) (Hilzheimer)
Leopard (*Felis pardus fusca*) (F.A.A Meyer)
Chinese small spotted tiger cat (*Felis bengalensis chinensis*) (Gray)

EDENTATES

Chinese pangolin (*Manis pentadactyla dalmanni*) (Sundevall).

RODENTS

Common house mouse (*Mus musculus*) (Linnaeus)
Brown rat or sewer rat (*Rattus norvegicus*) (Berkenhout).
Buff breasted rat (*Rattus rattus flavipectus*) (Milne-Edwards).
Smaller bandicoot rat (*Bandicota nemorivaga*) (Hodgeson).
Porcupine *Hystrix (Acanthion) subcrista subcrista*) (Swinhoe)

SIRENIANS

Dugong (*Dugong dugong*) (P.L.S.Muller)

UNGULATES

Reeves' muntjac or barking deer (*Muntiacus reevesi*) (Ogilvy)
Wild boar (*Sus Scrofa chirodonta*) (Heude)

Herklots admitted at the time that the list omitted all cetaceans and chiroptera. The list above is exactly as he wrote it as far as nomenclature is concerned. We realise that the latin names of several species have changed but in the interests of historical accuracy we have left them as Herklots named them. The only changes we have made are to include the otter with the mustelids and to include the porcupine with the rodents.

Herklots' list is far more useful than the 1898 reference and it can be seen that the animals are an interesting mixture of tropical and Palaearctic species. Only sixteen years later, the next list makes very different reading. Although it contains the same broad mix of species types the actual species vary tremendously:

INSECTIVORES

Large musk shrew or house shrew (*Suncus murinus*)
Woodland shrew (*Crocidura attenuata*)

PRIMATES

Long tailed macacque (*Macaca irus*)

CARNIVORES
Mustelids

Common ferret badger (*Melogale moschata*)
Eastern Chinese otter (*Lutra lutra chinensis*)

Canids

South China red fox (*Vulpes vulpes hole*)

Viverrids

Seven banded civet (*Viverricula indica*)
Chinese civet (*Viverra zibetha*)
Masked palm civet (*Paguma larvata*)
Crab eating mongoose)*Herpestes urva*)

Felids

Chinese leopard cat (*Felis bengalensis chinensis*)

EDENTATES

Chinese pangolin (*Manis pentadactyla*)

RODENTS

Common house mouse (*Mus musculus)*
Unnamed subspecies (*Mus musculus bactrianus*)
Norway rat (*Rattus norvegicus)*
Buff breasted rat (*Rattus rattus flavipectus)*
Sladen's roof rat (*Rattus rattus sladeni)*
Unnamed rat (*Rattus rattoides)*
Eastern spiny-haired rat (*Rattus huang)*
Bandicoot rat (*Bandicota indica nemorivaga)*
Chinese porcupine (*Hystrix hodgesoni)*

UNGULATES

Reeves' muntjac or barking deer (*Muntiacus reevesi)*
Wild boar (*Sus Scrofa chirodonta)*

The following species were also listed as animals that had been recorded from the Colony since 1910:

Dugong (*Dugong dugong)*
South China tiger (*Panthera tigris)*
Leopard (*Panthera pardus*)
Wild red dog (*Cuon alpinus)*
Eurasian badger (*Meles meles)*

A 1981 listing named the following mammal species as resident from Hong Kong:

INSECTIVORES

Large musk shrew or house shrew (*Suncus murinus)*
Woodland shrew (*Crocidura attenuata)*

PRIMATES

Rhesus monkey (*Macaca mulutta)*
Crab eating macacque (*Macaca fascicularis)*

CARNIVORES
Mustelids

Common ferret badger (*Melogale moschata*)
Eastern Chinese otter (*Lutra lutra chinensis*)

Canids

South China red fox (*Vulpes vulpes hoole*)

Viverrids

Small Indian civet (*Viverricula indica*)
Large Indian civet (*Viverra zibetha*)
Masked palm civet (*Paguma larvata*)
Crab eating mongoose (*Herpestes urva*)

Felids

Chinese leopard cat (*Felis bengalensis chinensis*)

EDENTATES

Chinese pangolin (*Manis pentadactyla*)

RODENTS

Asian house mouse (*Mus musculus castaneus*)
Field house mouse (*Mus musculus homourus*)
Common rat (*Rattus norvegicus*)
Black rat (*Rattus rattus*)
Sladen's rat (*Rattus sladeni*)
Eastern spiny-haired rat (*Rattus huang*)
Buff breasted Rat (*Rattus rattus flavipectus*)
Smaller bandicoot rat (*Bandicota indica*)
Porcupine (*Hystrix hodgesoni*)
Burmese squirrel (*Callosciurus pygerythrus styani*)
Thai squirrel (*Callosciurus flavimanus thai*)

UNGULATES

Reeves' muntjac or barking deer (*Muntiacus reevesi*)
Wild boar (*Sus Scrofa chirodonta*)

Hill and Phillips also noted that sixteen species of bats had been recorded from the colony, but qualified this statement by saying that several species were only known from single specimens.

One would not imagine that the wildlife of such a restricted area would change dramatically between 1981 and 1996, but to the surprise of the present authors it has.

Dr. Gary Ades wrote to Richard Muirhead on the 6th January 1996, in answer to a query about the alien big cat reported from Shing Mun during the 1960s, which is discussed further in chapter two. His letter included a surprising paragraph:

"Other rare or uncommon mammals which are spotted from time to time include the Small Indian Civet, Masked Palm Civet, Javan and Crab Eating Mongoose, Ferret Badger, Pangolin and Otter (at Mai Po wetland reserve), Barking Deer, Wild Boar and Chinese Porcupines are now fairly common in some areas of Hong Kong".

The crab eating mongoose, had been rediscovered several years before and an entirely new species of mongoose had been discovered living in the wilds of the New Territories.

Another entirely new species, the yellow throated marten *(M flavigula)* which had previously only been recorded as a food animal on market stalls was recorded from the region in the early 1990s, adding yet another new species to the Hong Kong list.

In 1992, Goodyer presented the most recent (at time of writing) species list, and again there are definite and important changes.

INSECTIVORES

Large Musk Shrew or House Shrew *Suncus murinus.*
Grey Shrew *Crocidura attenuata*

PRIMATES

Rhesus Monkey *Macaca mulutta*
Crab eating Macacque *Macaca fascicularis*
Japanese Macaque *Macaca fuscata*
Tibetan Macaque *Macaca thibetana*

CARNIVORES

Goodyer notes that both the Dhole and the Red Fox appear to have become extinct in the territory as is the Chinese Badger but notes the following species:

Mustelids

Common Ferret Badger *Melogale moschata*
Eastern Chinese Otter *Lutra lutra chinensis*

Viverrids

Small Indian Civet *Viverricula indica*
Masked Palm Civet *Paguma larvata*

He notes that the Large Indian Civet *Viverra zibetha* "*is probably now extinct in the colony*".

Mongooses

Crab eating mongoose (*Herpestes urva*)
Javan mongoose (*Herpestes javanicus*)

FELIDS

Chinese leopard cat (*Felis bengalensis chinensis*)

Goodyer also notes that large numbers of feral domestic cats *(F.catus)* and feral dogs are living in the colony.

EDENTATES

Chinese pangolin (*Manis pentadactyla*)

RODENTS

Asian house mouse (*Mus musculus castaneus*)
Field house mouse (*Mus musculus homourus*)
Ryukyu mouse (*Mus caroli*)
Common rat (*Rattus norvegicus*)
Roof rat (*Rattus rattus rattus*)
Sladen's rat (*Rattus koratensis* (Formerly *R.sladeni*)
Chestnut rat (*Niviventer fulvescens* (Formerly *Rattus huang*)
Lesser rice field rat (*Rattus losea*)
Buff breasted rat (*Rattus rattus flavipectus*)
Greater bandicoot rat (*Bandicota indica*)
Porcupine (*Hystrix brachyura* (Formerly *H. hodgesoni*)
Burmese squirrel (*Callosciurus pygerythrus styani*)
Belly banded squirrel (*Callosciurus erythraeus*)

He theorises that although *C.erythraeus* is the animal identified by Hill and Phillips as *C. flavimanus thai* , and that there are quite probably more than two species of tree squirrel presently living in Hong Kong.

He also writes:

"In addition Ground Squirrels are sometimes seen in Hong Kong and these are probably escaped pets".

It is not known to which species of ground squirrel he is referring, but in recent years the chipmunk (*Neotamius* spp) and the Siberian chipmunk (*Tamias sibericus*) have become popular pets across the world, and are likely to have become established in several alien locations worldwide. In conversation with the authors, Plymouth based zoologist Chris Moiser has theorised that these are amongst the animals most likely to become established as the next addition to the naturalised fauna of the United Kingdom, as opposed to Baker (1988) who theorised that the next two species to become established in the UK would be the wild boar (*Sus scrofa*) [which did] and the raccoon (*Procyon lotor*) [which didn't].

It is also, we feel, important to note the changes in the 'English' names given. Some of these changes seem to be without any real rhyme or reason. For example, the animal once known as the lesser bandicoot rat is now known as the greater bandicoot rat.

We are perfectly aware that these common names can differ wildly across the world; the animal known as the common toad, for example is *Bufo bufo* in England, *Bufo woodhousi* in the United States and *Bufo melanostictus* in Hong Kong, for example, and the bird known as the robin in North America is twice the size and a completely different colour than the bird with the same name known in Europe, but the appellatives 'greater' and 'lesser' imply a difference in size which would now, apparently, have miraculously been reversed.

UNGULATES

Reeves' muntjac or barking deer (*Muntiacus reevesi*)
Wild boar (*Sus Scrofa chirodonta*)

Goodyer notes twenty species of bat known from Hong Kong. We have not included any of the bat lists here for several reasons. Although, as has already been noted, several of the records of bats are known only from single specimens, and are therefore apparently anomalous, both bats, and to a certain extent cetaceans, because of their mobility, present a different set of habitat problems to the zoologist than land mammals and should be dealt with in a different manner.

The controversy over the taxonomy of the mongooses of Hong Kong, perhaps deserves to be examined in more detail. As a child one of my favourite stories in Kipling's *'The Jungle Book'* was *'Rikki Tikki Tavi'*, the story of the brave mongoose who saved a whole family of people from being killed by evil incarnate, personified in the form of an Indian hooded cobra.

Hooded cobras were, and are, common in Hong Kong, and much to the joy of the writer at eight years old, mongooses were also known from the territory.

By the time that the CFZ began their research into Hong Kong's mystery zoofauna it seemed that the one species of mongoose known from Hong Kong was sadly extinct, and that this charming and fascinating little carnivore would take up no more than a footnote in this *magnum opus*. The truth, as it always seems to turn out, was far more complicated, and took on so many bizarre aspects that the story deserves to be told in full!

According to Herklots, the only mongoose known from Hong Kong was:

"Crab Eating Mongoose, Herpestes urva (Hodgson)

This animal is probably very rare in the Colony but as it is nocturnal it is not possible to determine exactly how rare it is; I have only seen one on Hong Kong Island and none in the Territories.

This mongoose is smaller than any of the civets and its tail is broader at the base tapering towards the tip. The upper parts are buff, the long hairs each with a broad central black band and a white tip thus giving a hoary appearance. The hairs on the under parts gave less black. The tail is buffish-white. The lips, chin and throat are white or whitish; there may be white stripes from the corner of the mouth to the shoulder. Length of body 20 inches, tail 9 inches, total length 29 inches."

Writing sixteen years later, Patricia Marshall described the animal slightly differently. She said that this species was *"grey with many white tipped hairs"*, and a *"white neck stripe"*.

She also said that the crab eating mongoose was:

"Large for a mongoose" and *"relatively unafraid of man"*. She described an animal that *"hunts by day and night"*, and lives *"near water"*, especially *"near swamps or mountain streams"* where it lives on loaches and the territory's three species of freshwater crab. She said that it is *"short sighted and appears bold; it can be approached fairly easily"*.

Her measurements for the size of these animals were also different to those given by Herklots only sixteen years before. Whereas Herklots described an *animal "Length of body 20 inches, tail 9 inches, total length 29 inches"*, her figures covered a far greater range, allowing a head and body length of between 405 and 840mm (16-33 inches) and a tail of 220-336 mm (8.75 - 13.25 inches). Herklots' measurements are within the parameters allowed by Marshall, but it is, perhaps significant that the two sets of measurements are strikingly different!

Marshall did note, however that in her opinion this mongoose:

"probably no longer exists in the Colony. It was present on Hong Kong Island and in the New

Herpestes javanicus in Hong Kong

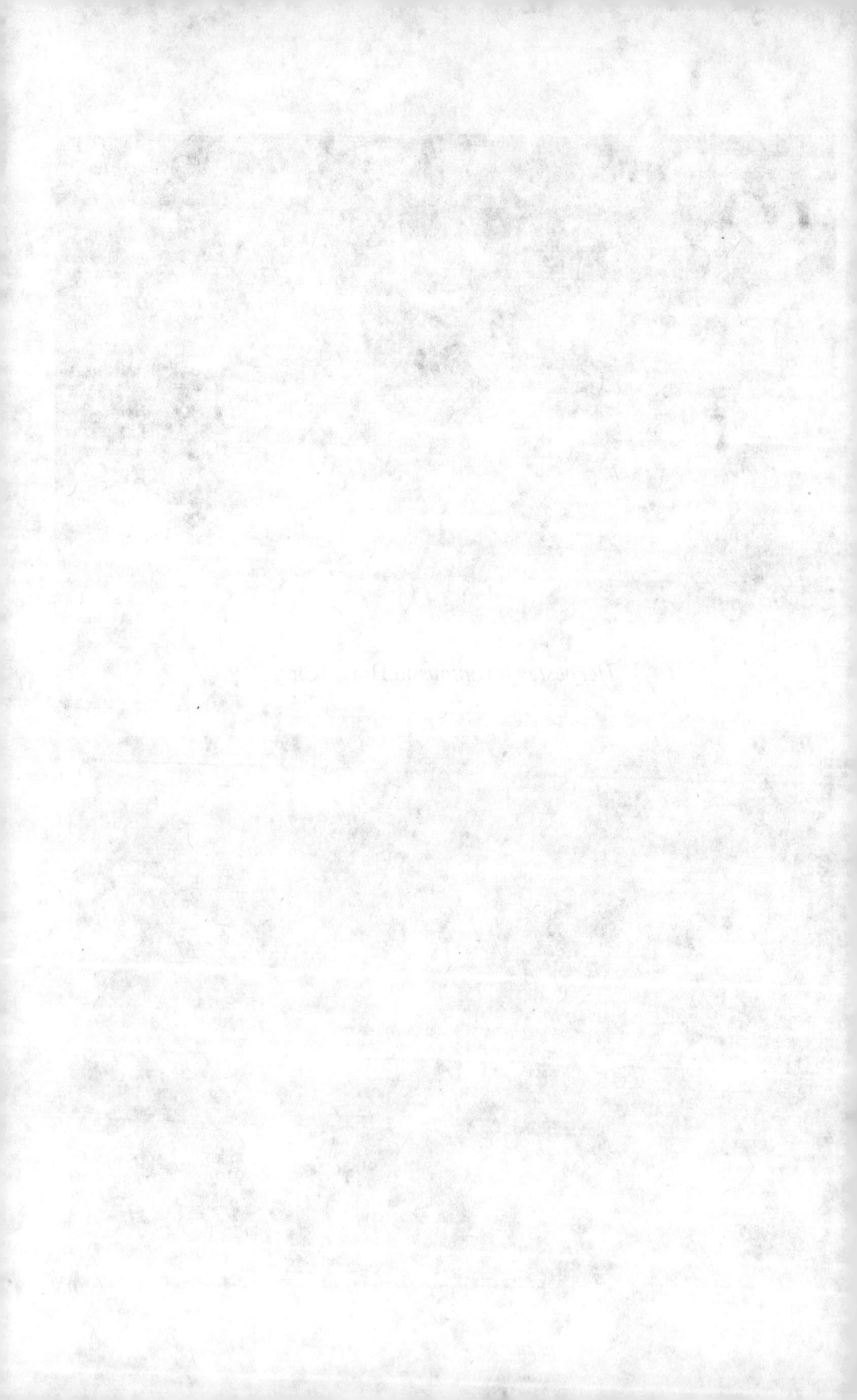

Territories until at least fifteen years ago".

It is unclear whether these records from the 1950`s, to which she referred are, in fact those of Herklots published in 1951. If they are, then it appears that they may well have been an unfortunate misinterpretation of what Herklots actually said. Although '*The Hong Kong Countryside*' was indeed published in 1951, much of it was written between 1941 and 1945 when Herklots was a prisoner of the occupying Japanese.

The sighting of a crab eating mongoose on Hong Kong Island to which Herklots refers, could well, therefore have taken place as early as the 1930s. It does appear to be the only firm record of this species from Hong Kong Island, and as such is quoted repeatedly throughout the available literature.

Hill and Phillips (1982) also claimed that:

"The last sightings of this species locally were probably in the early 1950`s and it is now thought to be extinct".

They give yet another set of measurements, which although like those given by Herklots are within the parameters given by Marshall, but unlike those of Herklots are in the upper range of Marshall's figures.
According to Hill and Phillips:

"It is a large mongoose, ranging from 80 - 110 cm in length and 1.5 - 2.2 kg in weight. It used to live near water, either mountain streams or marshy areas (freshwater), where it fed on rats, crabs, fish and various invertebrates. The claws are long and non retractile; five per foot. A very characteristic animal in appearance; it was apparently poor-sighted and could often be approached in the wild quite easily".

This description fits in well enough with that given by Marshall, but is so different from that given by Herklots that they could have been speaking of a different species.

Perhaps they were!

One of the first zoologists in Hong Kong to help the authors during their research was Dr Gary Ades of Kadoorie Farm, the Botanical Gardens and Conservation Centre in the New Territories. In January 1996 he wrote to Richard Muirhead listing some of the larger mammals that were still found in Hong Kong. As already noted this list included two species of mongoose. As far as the authors knew, and according to all the references cited above, which when the authors started work on this project were all they had available to them, there was only one species of mongoose in Hong Kong and this had been extinct since at least the early 1950s and if the author's suppositions about Marshall's misinterpretation of Herklots' data were correct, the animal had been extinct for much longer than that!

It appeared now that everything that had been written on the subject of mongooses in Hong Kong since at least 1951 was completely wrong. We had to re-evaluate the evidence in the light of this astonishing new discovery.

Goodyer (1992) listed the two species of mongoose in his Hong Kong mammal survey:

"This is another species that Marshall and Phillips (1965) feared was extinct. However it is still present in the Territory today. Crab-eating mongooses, as the name suggests, feed mainly on aquatic animals such as crabs, fish, molluscs etc.

They are large mongooses (head and body up to half a metre) and have an overall brown body colour made up of hairs that are individually ringed dark and light. There is a white patch on each side of the neck which may serve as warning colouration.

Most records come from the Mai Po/Tsim Bei Tsiu area with additional records from Sek King. Tei Mei Tuk and Ha Tseun just short of Tseun Mun. Herklots (1951) gave only one re-cord, which was from Hong Kong Island. There are no recent records from there. Mongooses are unusual amongst Hong Kong's mammals in that they are often seen to be active during the day".

It is interesting that Herklots' single record is the only historical record noted. Indeed as is be-coming clear, it is the only historical record of this species from the territory.

The next record is of a completely different species altogether. The Javan mongoose (*Herpestes javanicus*) (E.Geoffroy, 1818).

"The first record of the Javan Mongoose in Hong Kong was in Mai Po in May 1990 when one was accidentally trapped in a dog-trap, but it is suspected that this species was seen at Mai Po *two years previously. The identification of this species was verified by D.Wells, the regional authority on Mongooses on the basis of photographs, measurements and hair samples. Since that time several have been seen including adults with young. Mai Po is as yet the only place in Hong Kong where they have been seen. The Javan Mongoose is smaller than the Crab-eating Mongoose and has a brownish coat which is quite red around the head. When running, it has a squirrel-like motion with the body and tail close to the ground".*

It could perhaps be argued that because Herklots' measurements are so much smaller than any of the other three descriptions given, that Herklots' record may actually have been of *Herpestes javanicus* rather than *Herpestes urva*. It should also be questioned whether *Herpestes urva* actually became extinct in the territory. If we assume that the species did actu-ally become extinct then could it have become reintroduced from a mainland Chinese popula-tion, or perhaps from escapees from a food market.

Although, once it might have been argued that the population could have been replenished from a flourishing mainland population, when both sides of the border consisted of scrub land and paddy fields, by the late 1980's when the species was rediscovered, the situation was very different indeed.

Where once lay paddy fields now lie the concrete wastelands of the Shenzen Special Economic Zone, and it seems unlikely that any species of mongoose is resident there.

There is no doubt that several species of mongoose are regularly sold as food animals. A UK television series from the late 1980s included a clip of a specimen of *H. urva* in a distressingly small cage. Marshall (1967) notes that a third species of mongoose *Herpestes edwardsii* is also regularly seen. It does, however seem unlikely, especially given the physical condition of most wild caught `market mammals` that enough healthy adults of either species could have escaped into the surrounding environment to establish viable populations.

Here we should note that *Herpestes edwardsii*, the Indian mongoose, is of course *Rikki Tikki Tavi*, the fictional animal described by Kipling, which started off this investigation. This is the only reference that we have been able to unearth for this species, but Chan, Au and Young (1992) who described the discovery of the Javan mongoose in Hong Kong, revealed that the taxonomy of the Chinese mongooses as a whole was open to question.

They also revealed that not only was the actual date of discovery of this species from Hong Kong open to question, but that a specimen had been on open display in a Hong Kong museum before the animal was actually discovered from the territory.

Whereas Goodyer claimed that the first specimen had been caught in a dog trap in May 1990 Chan, Au and Young claimed that the first Hong Kong specimen was captured in a rattrap in November 1989. They obviously cannot both be right!

Chan, Au and Young added some convincing additional information:

"It seemed an incredible capture, for the trapping cage was barely larger than the animal and peanut butter was the bait!"

It is interesting to note that the story of the peanut butter, which does not seem particularly unlikely in itself re-surfaced several years later in 1996 when what appears to be a new species of rat was also said, according to a French news agency, to have been captured using traps baited with peanut butter. Either you have a group of colonial mammalogists with a penchant for peanut butter, or someone, probably in the French news agency is adding a touch of 'human interest' in an attempt to 'Disneyfie' an otherwise uninteresting story and make it `cute`.

In the mini museum at Pak Tam Chung Visitors Centre there was a stuffed animal labelled as a weasel. This in itself would have been particularly strange because (with one exception, which will be discussed later in this chapter), although there are several species of weasel and marten living in Guangdong the only member of the weasel family recorded from Hong Kong is the ferret badger (*Melogale moschata*).

This animal was in fact a mongoose, but it was not a Crab-eating mongoose. They note that there are only two species of mongoose known from southern China, the Crab-eating mongoose and the smaller `red cheeked` mongoose.

Chan, Au and Young include astonishing evidence which suggests that the precise identity of this mongoose, which is, presumably the same as the smaller species recorded from Mai Po, has variously been described as *Herpestes javanicus* and *Herpestes auropunctatus,* and although some authorities consider these two species to be conspecific, they do appear to be different species. They also quote Professor Wang who suggests that the only true specimens of *Herpestes auropunctatus* were from Yunnan. They also revealed that there appear to be two distinct sub-species of *Herpestes javanicus*, one from Yunnan and the other from Guangdong, Guangxi and Hainan.

This leaves a number of questions un-answered. The authors feel that they have effectively dismissed the suggestions that either species could have been introduced/re-introduced as a result of either natural population expansion from Guandong or as a result of escapees from market stalls. If we ignore the less scientific suppositions about species dispersal suggested by Charles Fort and his followers, and the authors, as fortean zoologists believe that as rational investigators we should not exclude any possibilities, we are left with two non-fortean models by which the introduction/re-introduction of these two species could have occurred.

These are that:

a) The crab-eating mongoose never became extinct and actually remained in viable numbers in Hong Kong between the second world war and 1990.

The Javan mongoose has always existed in the colony, and had managed to avoid discovery by such expert researchers as Herklots and Romer for a century and a half.

b) One or both species were introduced/re-introduced on purpose by someone who knew exactly what they were doing.

Introduction/re-introduction programmes have to be carried out with great care in order to ensure their success, and it seems eminently unlikely that a casual programme of introduction/re-introduction by a non-qualified person would have succeeded.

The other question that has to be addressed whilst considering this particular hypothesis is the question of where the hypothetical reintroduction programme had obtained their specimens. We have already seen that animals on sale for food in local markets often suffer from horrific injuries sustained during their capture and preparation for sale, and it seems eminently unlikely that such animals would be suitable stock for a breeding and introduction/re-introduction programme.

Chan, Au and Young noted:

"It seems that the animal was a real wild one because its condition was good and it showed no signs of being captured in a gin trap or of being kept in a a cage for a long time".

To suppose that such animals are purchased, nursed back to health and then released on a scale to allow a viable population to be established would, it seems, suggest the involvement of either a major animal welfare agency, such as the H.K.S.P.C.A or a major academic/zoological one on the scale of the University of Hong Kong or the Worldwide Fund for Nature. There is absolutely no evidence to suggest that any such organisation did anything of the sort, and despite the activities of other investigators working within the broad field of fortean and quasi fortean research, we are not in the habit of concocting elaborate conspiracy theories to prove the validity of our data.

The eyewitness accounts, which have been discovered since, suggest that both species are more widespread than one would otherwise have supposed.

According to Steve McChesney writing in the second issue of *'Porcupine!'* an excellent magazine published by the Ecology and Biodiversity department of the University of Hong Kong, during the late summer of 1992:

"An adult Javan mongoose (Herpestes javanicus), was often seen along the footpath from the Education Centre, with a party of from 1-3 young. The same thing happened during 1991 at around the same time of the year. An adult Javan mongoose has also often been seen sat the back of the Education Centre"

Chan, Au and Young noted the 1991 specimens from Mai Po as well:

"They were found along the border fence (one female was captured with a fish-net but released in April 1991), and near the Education Centre at Mai Po. One female and three cubs, only half the size of the female, were seen using the wall of the Mai Po Marshes Wildlife Education Centre as their den!

Apart from Mai Po, there were also other sightings of the same species. The earlier records of mongooses at Tai Po were probably this smaller species, and one was seen at Shalotung on the 27th October 1991"

Additional evidence for the theory that the discovery of *Herpestes javanicus* and the rediscovery of *Herpestes urva* from Hong Kong are part of a general resurgence of both species in southern China came when an article called 'The Oingping Report' was published in issue three of *Porcupine!* This collated lists of animals of various species seen on sale at the great Oingping food market at Guangzhou (Canton).

Whereas no mongooses had been seen on a visit in November 1990, four crab-eating Mongooses were seen in December 1991 and again, a year later, when eight specimens of *H. javanicus* were also seen. This suggests that perhaps both species are becoming more common, possibly as a result of a population explosion in the local rat population, but it must be stressed that these three records are too limited to be of any great statistical significance.

The same issue of *Porcupine!* noted a record by John Holmes of *H.urva* from the border fence at Tsim Bei Tsui on the 5th December 1992. Another animal of the same species was seen by Ms Phaik Hua Tan on the 20th February the following year also at Tsim Bei Tsui, confirming suggestions included earlier in this chapter that the species is well established there. On the 8th

April 1993, however, two Javan mongooses were seen at the same location by David Melville, suggesting that the two species are able to inhabit the same environment without any great competition for food.

On the 7th April 1993, a crab eating mongoose was seen about half a mile south of the Tsim Bei Tsui Police Post, on the Mong Tseung peninsula. According to George Walthew:

"It was walking beside a sludge filled pond, and was sandy grey in colour and finely grizzled with black".

He had three sightings of Javan mongooses that month:

"The first on the 13th April was approx. 2km along the fence from Tsim Bei Tsui - this specimen was a road kill. On 28th April I saw one animal about 200m along the fence from TBT. The third sighting was of two adults together at Nam Sang Wai".

On the 14th September the following year Graham Reels saw what *'may have been a Javan mongoose'* approximately 700 feet up Kun Yum Shan in the Kadoorie Farm site.

Steve Reels saw another unidentified mammal, which was probably a Javan mongoose, on the 9th April 1995. He described:

"A chestnut coloured, squirrel-like animal approximately 25-30 cm long. It had a long bushy tail (about half its length) and ran across the path about 10m in front of me. The sighting was made near Yuen Tun in Tai Lam Country Park".

As the editor of *Porcupine!* pointed out in a comment printed below this particular record, neither the Javan mongoose, or the other likely contender, a juvenile small Indian civet have a bushy tail as such.

Is this an example of an aberrant individual of either species? A misidentification? Or a true 'unknown animal'? The present authors would not like to hazard an opinion based on such flimsy data.

The previous two sightings, whilst unconfirmed, do suggest that mongooses of one or other species are increasing their range quite fast.

Another item in Porcupine! number 13, by Graham Reels confirmed this supposition:

"On 17th July I was walking through grassland at about six hundred metres near Kadoorie Farm when I came across two mongoose kittens apparently playimng on the footpath. They were approximately 25 cm nose-tail (tail approximately 10 cm) and had uniform dark brown fur. I am unsure as to the species. Of the two mongoose species known from Hong Kong; Javan Mongoose (Herpestes javanicus) and crab-eating Mongoose (H.urva), the latter, which is rarer is more likely to be found at such an altitude, and has dark brown fur. However, I caught a glimpse of what appeared to be an adult Javan Mongoose very close to this site sev-

eral months ago. Javan Mongoose are quite commonly seen in the lowlands of the northwest New Territories".

One would hardly credit that this animal which *is "quite commonly seen in the lowlands of the northwest New Territories"* is of the same species whose advent was treated with so much excitement only three or four years previously.

Our final record of 1995 comes from the same publication, and not surprisingly also took place at Tsim Bei Tsui. The short article was titled `*Tales of a Mongoose*` and was written by Andrew Cornish:

"Between April and May my walks along the border fence at Tsim Bei Tsui were frequently enlivened by mongoose crossing the road. These occurred at all times of the day and along some very different stretches of road so I do not believe that they were all the same animal (s).

Once I saw a pair but always the animals were moving to or from the particular large bunds covered with Phragmites that line the road in places. As I never saw a mongoose emerge from the fishpond side of these bunds it seems likely that they have burrows in them. these bunds are certainly the closest point to the mangrove the mongoose could burrow without constant fear of flooding. All the animals I got close enough to identify were Javan Mongoose (Herpestes javanicus)".

It seems from analysis of the above evidence, that whilst Javan mongooses only are found at Mai Po Marshes, both species are found at Tsim Bei Tsui and possibly at Kadoorie Farm. Whilst it is undeniable that many sightings have taken place along the border fence, we feel it unlikely that these animals could have been all recent immigrants from China.

Until relatively recently the border fence area was a restricted zone, and it seems more likely that both species have co-existed in an area not usually visited by humans, undetected.

When humans, especially those with an interest in the natural sciences started to visit the area the animals were `discovered`. This, however does not even start to explain the resurgence of mongoose activity elsewhere in the New Territories. For the moment it must remain a mystery.

The most recent zoological mystery from Hong Kong concerns the precise identity of the Barking Deer. Although., as we have seen these creatures have been known from Hong Kong for centuries, it has always been stated that they are Reeve's muntjac *M. reevesi*.

However evidence presented during the part three years has suggested that some at least of the Hong Kong deer are actually Indian muntjac (*M muntiacus*) – a slightly larger species and at the time of going to press, no-one seems prepared to go on the record and state exactly which species of deer are living where in Hong Kong.

The examples outlined above are only a few of the dozens of examples of the peculiar history of Hong Kong`s zoology which we have on our files. The project has been ongoing since the

CFZ was started and shows no sign of abating.

The data in the lists presented above are to be analysed in more depth in the ongoing CFZ Hong Kong Project, of which this paper is only a part. However, what is certain is that the wildlife of the region is as changeable as its people or politics, and that it is a foolish man indeed, even in the most densely populated place on earth, who even attempts to produce a definitive listing or to say that there is nothing left to be discovered there.

REFERENCES

GOODYER, N. `Notes on the Land Mammals of Hong Kong` (Memoirs of the Hong Kong Natural History Society, 1992).
MARSHALL, P. `Wild Mammals of Hong Kong` (London, OUP 1967)
BROWN, J., and LEE, H.,`Hong Kong and Macao - the rough guide` (Rough Guides, London 1993)
HUGHES, R `Hong Kong - Borrowed Place - Borrowed Time` (London, Deutch 1968)
HEUVELMANS, Dr. B. `On the track of Unknown Animals` (First English Language Edition, London, Rupert Hart-Davis 1958).
MORSE `Chronicles of the East India Company Trading to China (Chronicles)`, Vol i, p.32
WELSH, Frank `A History of Hong Kong` (London, Harper Collins, 1993)
WILSON, D., `Hong Kong! Hong Kong!` (London,Unwin Hyatt, 1990)
HERKLOTS, G.A.C., `The Hong Kong Countryside` (Hong Kong, SCMP 1951)
HILL D.S., & PHILLIPS, K., `A Colour Guide to Hong Kong Animals` (Hong Kong, Government Publications Office 1981).
ADES Dr. G. Pers Corr to Richard Muirhead 6.1.96
LETHBRIDGE, H.J., (introduction) `The Hong Kong Guide 1893` (Hong Kong, University Press reprint 1982)
HERKLOTS, G.A.C., `Hong Kong Birds` (Hong Kong SCMP, 1968)
GILLETT, K.E. `The Chipmunk and the Siberian Chipmunk in Captivity` (Plymouth, Bassett, 1988).
SMITH H.M, & BARLOWE, S. `A Guide to Field Identification - Amphibians of North America` (New York, Golden 1978)

"Cry wolf": The mystery killer of the Australian Outback

by Richard Freeman

Patterns in nature repeat themselves. A good design will appear again and again. This is called convergent, or parallel, evolution. Where environmental pressures are alike in different areas of the world, totally unrelated creatures will evolve to resemble each other. This may occur in differing time periods as is the case with ichthyosaurs (fish eating marine reptiles of the Mesozoic era) and modern dolphins, or in the same time period as in old world vultures and new world vultures (that are more closely related to storks!)

The most spectacular example of convergent evolution is that of *Thylacinus cynocephalus;* the Tasmanian wolf. This animal is a carnivorous marsupial related to the Tasmanian devil. Approximately the size of a placental wolf the thylacine's resemblance to its namesake is sticking. It has a dog like head, but its toothy jaws can open far wider than that of a true canid, up to 120 degrees. The body is similar but with shorter legs. Though not as fast as a real wolf, the thylacine has much more stamina and can run down its prey over many miles. The hind legs resemble those of a kangaroo and the creature can rear erect and leap in the famous marsupial fashion. The tail is long and stiff to counter balance the body. It does not taper like a dog's tail and is not used for signalling. Its short coat is veritable in colour from grey, to yellowish, to brown. The thylacine's most distinctive markings, however, are a set of stripes running from the shoulder to the rump. These are dark brown or black and can vary in number between 13 and 22, females generally having more than males. These markings have led to the animal sometimes being referred to as the Tasmanian tiger. This is a pity, as it confuses the issue, and in reality the creature is convergent with the dog family not the cats. The thylacine is unusual in marsupials as both sexes bear a pouch. The pouch points backwards to prevent entanglement in the bush.

Fossil evidence shows that the thylacine once lived in New Guinea. It was present on mainland Australia until around 3000 years ago. The introduction of the dingo seems to have lead to their disappearance on the mainland. Though not as physically formidable as the thylacine, the placental dingo could reproduce faster and may have also transmitted diseases. Dingoes, however, did not reach the island of Tasmania.

This last haven of the marsupial wolf was discovered by Dutch mariners in 1642 and was called Van Diemen's land until 1854. The first British settlement was established in 1803, and Tasmania soon became one of the most appalling penal colonies in the British Empire. With the white man came a familiar wave of destruction. The Tasmanian Aborigines were hunted like vermin by 1876 they had been wiped out. The black Tasmanian emu suffered a similar fate. The settlers began sheep farming, their docile, slow charges proved irresistible prey to the thylacine, and inevitable conflicted followed.

Thylacines were shot on site from the 1820. This escalated in 1830 when the Van Diemen's land company, a large pastoral firm began to offer a bounty foe each wolf skin brought in. By 1888, farmers were whining that the wolves were killing 40,000 sheep annually, a ridiculously high figure. The government caved in to these bleatings and offered their own bounty of one pound per wolf carcass. Over the next 21 years, 2898 bounties were paid as thylacine hunting became profitable.

Then in 1905 a wave of distemper, passed on by the settlers' dogs, swept through the already dwindling thylacine population. The numbers crashed. Spinelessly, the government allowed the hunting of thylacines to continue. In 1933 what was believed to be the last wild thylacine was captured in the Florentine valley. Some film was taken of Benjamin shortly before his death. This sad film shows him pacing about his cage and occasionally displaying his spectacular yawn. This magnificent creature was the last of his kind on earth, destroyed by human pig ignorance and greed - or was he?

Tasmania is larger than Scotland and has a tiny population concentrated mainly in the south east of the island. Much of Tasmania is mountain and Forrest, the latter some of the most dense in the world. Incredibly, large areas of Tasmania still have not been accurately mapped. Since 1936, there have been hundreds of sightings of living Thylacines in the Tasmanian bush.

In 1937, a large portion of south western Tasmania was set aside as a wildlife reserve and the following year an expedition travelled in to the area to search for evidence of thylacines. The renowned Bushman Arthur Fleming reported finding the animal's distinctive tracks one of which he made a cast of. This led leading Tasmanian naturalist Michael Sharland to conclude that despite reduced numbers the creatures were still widespread over southwestern Tasmania. Live traps were set for thylacines in 1945 and 1946. No specimens were captured, but tracks were found outside the traps. Adye Jordan was employed by the naturalist Sir Edward Hallstrom to trap thylacines in order to start a captive breeding programme. The Tasmanian Fauna Board cancelled the trapping permit in February 1949. Jordan claims he caught a female wolf twice in his traps, the first time just six days after the permit had been cancelled. Both times he had to let her go.

In 1944 Mick Tiffin a road patrol man for the Cardigan river was engaged in sawing up a fallen tree for fire wood. He was startled when a young thylacine rushed out of the hollow and away in to the scrub with an alarming cry.

Two men both saw a thylacine in December 1947. The animal was described as grizzled grey with stripes down the lower part of its back. Witnesses B. Thorpe and A. Woolley observed the beast in broad daylight at 7 am. It was chasing a wallaby and ran within 20 yards of them. Between 1936 and 1950 there were 36 reports, the numbers grew as time went on. According to a 1980 report by the National Parks and Wildlife Service 315 sightings were on record. The number now stands at well over 400. These have included some impressive witnesses over the years. Charles Abel was Tasmania's most famous bushman and had observed wolves many times in his youth. In 1955, he reported seeing a female with two cubs.

In 1982, Hans Naarding a wildlife researcher studying Latham's snipe - a small bird - was sleeping in his landcruiser near the head waters of the Salmon River, north west Tasmania on the evening of March 9. He awoke at 2 am and shone his spotlight around. It came to land on an incredible creature.

"I saw the animal for three minutes, which is one hell of a long time. It stood absolutely still and every part of him was visible. It was a fully-grown male wearing a fine (sandy-coloured) coat which was in good condition counted twelve black stripes over its back. It had a massive angular head and small rounded ears. The tail was very slender, but very thick at the butt, quite unlike the tail implant of a dog"

The creature gave a diagnostic wide yawn. When Naarding moved to get his camera the animal noticed him and slipped away like a ghost into the night. It left behind a strange musky odour.

This sighting is important for two reasons. Firstly Naarding was a trained zoologist who viewed the animal from only 20 feet for a long period, he is unlikely to have been mistaken. Secondly, there were many other sightings in the area in 1982. Park ranger Nick Mooney, who examines claimed sightings judged these to be "excellent".

Another witness unlikely to be in error is Turk Porteus. In 1929 he caught a female thylacine and her three cubs and sold them to Hobart Zoo after caring for them for three weeks. 53 years later he came face to face with another specimen. Whilst walking near the Frankland River in 1986 he heard a strange rustling in the scrub. Upon investigation he found a female thylacine only 60 feet from him. It was bluish-grey with 16 stripes. The loosely hanging pouch indicated she had recently raised cubs. Later he found her tracks and those of two young.

In 1995, park ranger Charlie Beasely watched a thylacine from around 500 feet in north east Tasmania, inland from St Helens. These men live and work in the Tasmanian bush and know what they are looking at. The encounters described hear are only a fraction of the reported number. Many more must go unreported due to fear of ridicule. The sheer amount of reports, the fact that so many include sightings of females with cubs, and the unimpeachable nature of some of the witnesses the case for the thylacine's survival on Tasmania seems very good.

One last piece of evidence comes from Dr Henry Nix of the Australian National University. Using a computer programme called BIOCLIM he plotted the thylacine's known preferences in habitat and prey against the changes in Tasmania's vegetation and demography. He then instructed the programme to create a detailed map of Tasmania highlighting the areas most favourable to thylacines if they still existed today. Finally, from the huge list of thylacine sightings from the past 60 years he chose those those that seemed reliable and compared the areas they took place in to the areas predicted by BIOCLIM. The match was almost exact. The chances of such a conformity happening at random is zero. Dr Nix announced his results in March 1990 concluding that the witnesses were really seeing thylacines, and that in undisturbed regions of their former range they may now be no rarer than they were prior to European innovation.

Thylacines surviving on Tasmania would come as no surprise to scientists but what of specimens on *mainland* Australia were the animal is supposed to have died out 3000 years ago? Incredibly there are even more reports of thylacine sightings from the mainland than there are from Tasmania!

It would be pointless to try and list the legion of mainland sightings that have occurred all over the vast land of Australia so I will recount some of the best from each area.

Mr Sid Slee of Yoongarllup, Western Australia has dubbed his farm "Hillside - the haunt of the marsupial wolf". He claims to have seen many specimens on his land since the 1940s. They apparently mauled 7 heifers and have killed many kangaroos whose carcasses he has come across over the years. His closest encounter occurred in 1972 when a massive thylacine emerged from the bushes right in front of him as he was doing his morning rounds. It was 7 feet long, sandy coloured, with chocolate brown stripes. The beast was so close to Sid that when it turned to run it kicked gravel over his boots. The kangaroo corpses he examined have generally been headless. Tasmanian farmers were said to have reported that thylacines were fond of biting into sheep snouts to rip open blood vessels and lap up the blood. One of these bodies was photographed and shows a long thylacine like print close to the carcass. Sid has taken convincing casts of several such prints.

In 1969 Archie Anderson, a Western Australia police officer and keen wildlife photographer was driving near Canarvon with his wife, and another couple and their daughter. At a distance of only 5 metres a thylacine appeared by the roadside. As the animal bounded away Archie shot some film with his cine camera. The pictures are not clear, and are at a distance of 90 feet, but seem to show a dog like animal with heavy hind quarters. Wildlife expert Ian Offer believed the film to be the best evidence that the thylacine still existed in Western Australia. He was impressed with the animal's gait, loping with the fore and hind legs not co-ordinated, typically marsupial.

If anywhere could hide a population of marsupial wolves it is the jungle-shrouded wilds of Queensland. Here Percy Trezise, a bush pilot, artist, and rock art expert has his remote home close to the Little Laura River. One July night in 1993, a bizarre animal was illuminated in the headlights of his land cruiser. The animal was striped with hindquarters resembling a kangaroo's. Trezise was familiar with dingoes having even raised one himself. Since then he has found their tracks, larger and deeper than dingo tracks, and imprints of the diagnostic stiff tail. He is convinced that a group of thylacines are living in the area.

Some areas have so many reports that the resident animals are given names in much the same way feral big cats are in Britain. One such is the Wonthaggi monster that has been devouring sheep and frightening witnesses in Southern Victoria since 1955. Well over a hundred people claim to have seen thylacines here.

Typical was the sighting of Ern Featherstone, a car salesman, who was demon-strating a vehicle to Mr. and Mrs. T. J.Schmedje in December 1955. They had been discussing the "monster" when it appeared from the scrub and ran along-side of the road. It was sleek, brown striped and reminded Mr Schmedje of a "hyena with a tail" and moved like a wallaby does when running on all fours. This area has, to date, produced more thylacine reports than anywhere else in the world.

Close by is the territory of the Orzenkadnook tiger on the Victoria / South Australia border. Here too have been many sightings since 1885, but also a photograph. Rilla Martin, a Melbourne woman on holiday in the area photographed a bizarre animal briefly seen in bushes beside the road. The unclear, yet intriguing, picture shows a thylacine-shaped animal partially obscured by vegetation. The head and the slope of the animal's back seem very like a thylacine. The stripes however seem to run up as far as the animal's neck unlike a thylacine that is striped on the hind quitters. It has been suggested that these "stripes" are merely shadows thrown upon the animal from the bushes. An examination of the negative revealed no trickery.

Most amazingly, and perhaps most believably, thylacine reports have recently emerged from the mountains of Iran Jaya in New Guinea, one of the least explored places on Earth. Tasmanian thylacine researcher Ned Terry received a letter from a retired missionary that recounted him showing natives pictures of Australian animals. They recognised the thylacine and said it inhabited the mountains. They called it the "dobsenga". Their description said it had a head and forelegs like the white man's dog, and a slim middle that bore stripes, and a long stiff tail.

Ned investigated in person. He was told that the dobsengas imaged from the mountains occasionally to devour pigs, chickens and other domestic animals. The people feared the dobsenga and did not hunt it. He ventured into the mountains but saw very little wildlife. The natives told him that if the prey animals such as cuscus, wallabies, and possums were not present the dobsenga would not be there. The animals followed their prey's movement. Ned intends to return.

The thylacine has been described as the world's healthiest extinct animal and with good reason. As the years go by sightings mount up and it is now only a matter of time before irrefutable evidence of this magnificent animal's continued survival is obtained. This is an animal that may have been given a second chance. Mankind too may have a second chance, to protect this living fossil - perhaps the greatest symbol of the world in our hands. In the words of my fellow cryptozoologist, Karl Shuker, in his excellent book *The Lost Ark*.....

"Let us all hope very sincerely that everything possible will be done to ensure that this time Tasmania's most splendid animal does not suffer the same terrible fate at the hands of mankind that it experienced during its previous period of "official" existence.

1999-2000
SPANNING THE MILLENIUM
WITH THE CENTRE FOR
FORTEAN ZOOLOGY

The twenty-four months ending at New Year's Eve 2000 were momentous ones within the Centre for Fortean Zoology. All three of the core members of the faculty had considerable successes - and failures - in both their professional and personal lives, and for much of the time it seemed as though the Good Ship CFZ was adrift on a stormier sea than usual.

1999 started fairly normally. My book, *The Rising of the Moon* – the third part of my trilogy exploring the links between zooform phenomena and other anomalous fortean happenings was published in April and was, much to my surprise, somewhat of a runaway success. However, almost from the start relations between the publishers and me were fraught and we parted company later that year with a certain amount of acrimony.

The second book in the trilogy, *Only Fools and Goatsuckers* has just been published by CFZ Press and new editions of *The Owlman and Others* (the first of the three) and *The Rising of the Moon* will be released as soon as stocks of the previous editions are exhausted.

At the 1999 Unconvention we made friends with Tony Healey, the veteran Australian cryptozoologist and co-author of the book *Out of the Shadows"*. A casual invitation from us to him that he "must pop down for a few days" ended up with him living in a van in the CFZ car park for the best part of three months. He is a lovely bloke and we all missed him greatly when he flew back to the Antipodes.

In July 1999, I was appointed editor of *Quest* Magazine. This was somewhat of a double edged sword as the management of the company employing me turned out to be deeply unsound, and my liaison with the company only resulted in one issue of the magazine being issued and a lot of good friends and colleagues of mine being owed an awful lot of money. Here I would like to apologise publicly to Karl Shuker, Lionel Fanthorpe, Nick Redfern, Dave Walsh, Graham Inglis and Richard Freeman amongst others for having got them mixed up with a company this dodgy! Just remember guys that revenge is a dish best served cold.

However, one of the few benefits of working on *Quest* was that we made friends with several nice people, some of whom, like Tim Matthews and Mike Hallowell remain valued colleagues to this day.

In the summer of 1999 we accompanied Uri Geller, et al on a delightfully boozy trip to the middle of the English Channel to see the total eclipse of the sun, and a few days later I flew to

Nevada where I was booked to appear at the International UFO Congress. Amongst the interesting people I met there was Dr Lloyd Pye – the custodian of the infamous `Starchild Skull` - a specimen that many believe to be a fake, and some believe to be the *bona fide* product of a mating between a human being and an extra terrestrial. I have no idea what it is, except for the fact that it is manifestly genuine, and that as I don't believe in aliens it has to be something earthbound in nature. We await more of Lloyd's results with interest.

We ended 1999 by being appointed staff writers on what turned out to be a short lived venture – *The Planet on Sunday* – and as the new century dawned things were looking quite bright for us all.

Unfortunately, the newspaper closed down in early February after only seven issues and all three of the faculty members were left on our uppers. We had used much of the money from our short-lived foray into the weekly press to upgrade the infrastructure of the CFZ, and to repair and upgrade cages for our menagerie.

We were now left with time to pursue our researches and no money with which to do it. CFZ Press published two new books by me in the first part of the year, and two other books, including *Weird Devon* - the official casebook of the Exeter Strange Phenomena Research Group (the sister organisation of the CFZ) - were published by Bossiney Books in the spring.

In April, the long awaited CD *The Weird World of Lionel Fanthorpe, Jon Downes and the Amphibians from Outer Space* was finally released on Voiceprint records. Nobody bought it, but to date it is probably the best punk rock crypto album ever made!!!!!

Also in May came the premiere of our ridiculous movie *The Owlman and Others*. From an early age I had wanted to be a film director, and now, finally here was my chance. We presented the gobsmacked Unconvention audience with a ridiculous farrago of nonsense featuring nudity, sham violence and gay nazi's which should, or so various pundits claimed, provide a beacon for all other wannabe fortean film makers for years to come.

In May, we presented our first annual conference *The Weird Weekend* with some success. We played host to David Farrant (The Highgate Vampire), Mike Hallowell (The Mystery of Marsden Grotto), Nick Redfern (The CIA Files on Noah`s Ark), Emmet Sweeney (Revisionist Biblical Archaeology), and Malcolm Robinson (Poltergeists of Scotland).

In June, my dearest friend, Toby the CFZ Dog died of cancer. He had been my constant companion for sixteen years and I still miss him terribly. A few months later my fiancée and I split up and I was thrown into a spiral of illness that lasted for the rest of the year.

Toby wasn't the only one of our animals to die during 2000. Isabella the little black cat (1985-2000) died a month later and was soon followed by Carruthers the grey cat (1986-2000) and at the end of the year Flump the Chinese soft shelled turtle (1990-2000), Cuddles the amphiuma (1991-2000) and even Steve the axolotl (1998-2000). They will all be sadly missed.

In September I visited South Wales to investigate the so-called *Beast of Trellech*. This was where a big cat had allegedly attacked a small boy. The laws of libel and slander do not allow me to print all my findings, but sufficient to say that I unearthed no evidence to suggest that anything of the sort had occurred.

The following month Richard and I were guests on *The Big Breakfast* where Richard swore, I played word games with Johnny Vaughn and made a joke about heroin. However the best is yet to come.

The high spot of the last few years was Richard's expedition to northern Thailand in search of a semi legendary dragon called the Naga. His adventures, and tentative conclusions are written up in full in *Animals & Men #23*. However this was undoubtedly the greatest CFZ achievement at least since Graham's and my expedition to Central America in 1998, if not ever. It has certainly given us new heights to aim for as we progress into the 21st Century.

Back at home things were not going well, and our spiral of illness and bad vibes culminated when Tracey Freestone, a friend of us all and sometime CFZ cohort committed suicide.

However, she left us to look after Tessie, her seven year old collie cross bitch, and from the moment that Tessie became an integral part of CFZ life things began to improve. Maybe the CFZ can't function without a CFZ doggie, but whatever the reason, so far at least 2001 is beginning to look up and I feel certain that when I write my next review of the year in the next yearbook, I will be painting a much brighter picture.

Slainte Mhor

THE CURRENT CREW OF THE CFZ MOTH-ERSHIP ARE:

Director: Jonathan Downes
Deputy Director: Graham Inglis
Zoologist: Richard Freeman
Magazine cartoonist and artwork: Mark North
Associate founding editor: Jan William
Tessie the CFZ Dog
Joyce Howarth and Alyson Diffey: Hedgewitches
Lisa Peach and Maxine Pearson do their own inimitable thing

CONSULTANTS

Consulting Editor: Dr Bernard Heuvelmans

Cryptozoology: Dr Karl Shuker, Dr Lars
Thomas, Loren Coleman
Zoology: Chris Moiser
Cetology and Palaentology: Darren Naish
Surrealchemist in Residence: Tony "Doc" Shiels (yes folks he is back)

REGIONAL REPRESENTATIVES

Scotland: Tom Anderson
Surrey: Nick Smith
West Midlands: Dr Karl Shuker
Kent: Neil Arnold
Gtr Manchester & Cheshire: Allen E Munro
Hampshire: Darren Naish
Leicestershire: Alistair Curzon
Cumbria: Brian Goodwin
Yorkshire: Steve Jones
Tyneside: Mike Hallowell

USA: Loren Coleman
Denmark: Dr Lars Thomas, Erik Sorensen
Northern Ireland: Gary Cunningham
Republic of Ireland: Daev Walsh
Spain: Alberto Lopez Acha, Angel Morant Fores
Germany: Hermann Reichenbach, Wolfgang
Schmidt
France: Francois de Sarre
Mexico: Dr R A Lara Palmeros

THE CENTRE FOR FORTEAN ZOOLOGY

So, what is the Centre for Fortean Zoology?

We are a non profit-making organisation founded in 1992 with the aim of being a clearing house for information, and coordinating research into mystery animals around the world. We also study out of place animals, rare and aberrant animal behaviour, and Zooform Phenomena; little-understood "things" that appear to be animals, but which are in fact nothing of the sort, and not even alive (at least in the way we understand the term).

Why should I join the Centre for Fortean Zoology?

Not only are we the biggest organisation of our type in the world, but - or so we like to think - we are the best. We are certainly the only truly global Cryptozoological research organisation, and we carry out our investigations using a strictly scientific set of guidelines. We are expanding all the time and looking to recruit new members to help us in our research into mysterious animals and strange creatures across the globe. Why should you join us? Because, if you are genuinely interested in trying to solve the last great mysteries of Mother Nature, there is nobody better than us with whom to do it.

What do I get if I join the Centre for Fortean Zoology?

For £12 a year, you get a four-issue subscription to our journal *Animals & Men*. Each issue contains 60 pages packed with news, articles, letters, research papers, field reports, and even a gossip column! The magazine is A5 in format with a full colour cover. You also have access to one of the world's largest collections of resource material dealing with cryptozoology and allied disciplines, and people from the CFZ membership regularly take part in fieldwork and expeditions around the world.

How is the Centre for Fortean Zoology organized?

The CFZ is managed by a three-man board of trustees, with a non-profit making trust registered with HM Government Stamp Office. The board of trustees is supported by a Permanent Directorate of full and part-time staff, and advised by a Consultancy Board of specialists - many of whom who are world-renowned experts in their particular field. We have regional representatives across the UK, the USA, and many other parts of the world, and are affiliated with other organisations whose aims and protocols mirror our own.

I am new to the subject, and although I am interested I have little practical knowledge. I don't want to feel out of my depth. What should I do?

Don't worry. We were *all* beginners once. You'll find that the people at the CFZ are friendly and approachable. We have a thriving forum on the website which is the hub of an ever-growing electronic community. You will soon find your feet. Many members of the CFZ Permanent Directorate started off as ordinary members, and now work full-time chasing monsters around the world.

I have an idea for a project which isn't on your website. What do I do?

Write to us, e-mail us, or telephone us. The list of future projects on the website is not exhaustive. If you have a good idea for an investigation, please tell us. We may well be able to help.

How do I go on an expedition?

We are always looking for volunteers to join us. If you see a project that interests you, do not hesitate to get in touch with us. Under certain circumstances we can help provide funding for your trip. If you look on the future projects section of the website, you can see some of the projects that we have pencilled in for the next few years.

In 2003 and 2004 we sent three-man expeditions to Sumatra looking for Orang-Pendek - a semi-legendary bipedal ape. The same three went to Mongolia in 2005. All three members started off merely subscribers to the CFZ magazine.

Next time it could be you!

Project Kerinci, Sumatra - 2003
In search of the bipedal ape Orang Pendek

How is the Centre for Fortean Zoology funded?

We have no magic sources of income. All our funds come from donations, membership fees, works that we do for TV, radio or magazines, and sales of our publications and merchandise. We are always looking for corporate sponsorship, and other sources of revenue. If you have any ideas for fund-raising please let us know. However, unlike other cryptozoological organisations in the past, we do not live in an intellectual ivory tower. We are not afraid to get our hands dirty, and furthermore we are not one of those organisations where the membership have to raise money so that a privileged few can go on expensive foreign trips. Our research teams both in the UK and abroad, consist of a mixture of experienced and inexperienced personnel. We are truly a community, and work on the premise that the benefits of CFZ membership are open to all.

What do you do with the data you gather from your investigations and expeditions?

Reports of our investigations are published on our website as soon as they are available. Preliminary reports are posted within days of the project finishing.

Each year we publish a 200 page yearbook containing research papers and expedition reports too long to be printed in the journal. We freely circulate our information to anybody who asks for it.

No. Each year since 2000 we have held our annual convention - the *Weird Weekend* - in Exeter. It is three days of lectures, workshops, and excursions. But most importantly it is a chance for members of the CFZ to meet each other, and to talk with the members of the permanent directorate in a relaxed and informal setting and preferably with a pint of beer in one hand. Since 2006 - the *Weird Weekend* has been bigger and better and held in the idyllic rural location of Woolsery in North Devon. The 2008 event will be held over the weekend 15-17 August.

Since relocating to North Devon in 2005 we have become ever more closely involved with other community organisations, and we hope that this trend will continue. We also work closely with Police Forces across the UK as consultants for animal mutilation cases, and we intend to forge closer links with the coastguard and other community services. We want to work closely with those who regularly travel into the Bristol Channel, so that if the recent trend of exotic animal visitors to our coastal waters continues, we can be out there as soon as possible.

We are building a Visitor's Centre in rural North Devon. This will not be open to the general public, but will provide a museum, a library and an educational resource for our members (currently over 400) across the globe. We are also planning a youth organisation which will involve children and young people in our activities. We work closely with *Tropiquaria* - a small zoo in north Somerset, and have several exciting conservation projects planned.

Apart from having been the only Fortean Zoological organisation in the world to have consistently published material on all aspects of the subject for over a decade, we have achieved the following concrete results:

- Disproved the myth relating to the headless so-called sea-serpent carcass of Durgan beach in Cornwall 1975
- Disproved the story of the 1988 puma skull of Lustleigh Cleave
- Carried out the only in-depth research ever into the mythos of the Cornish Owlman
- Made the first records of a tropical species of lamprey
- Made the first records of a luminous cave gnat larva in Thailand.
- Discovered a possible new species of British mammal - the beech marten.
- In 1994-6 carried out the first archival fortean zoological survey of Hong Kong.
- In the year 2000, CFZ theories where confirmed when an entirely new species of lizard was found resident in Britain.
- Identified the monster of Martin Mere in Lancashire as a giant wels catfish
- Expanded the known range of Armitage's skink in the Gambia by 80%
- Obtained photographic evidence of the remains of Europe's largest known pike
- Carried out the first ever in-depth study of the *ninki-nanka*
- Carried out the first attempt to breed Puerto Rican cave snails in captivity
- Were the first European explorers to visit the `lost valley` in Sumatra
- Published the first ever evidence for a new tribe of pygmies in Guyana
- Published the first evidence for a new species of caiman in Guyana

EXPEDITIONS & INVESTIGATIONS TO DATE INCLUDE:

- 1998 Puerto Rico, Florida, Mexico *(Chupacabras)*
- 1999 Nevada *(Bigfoot)*
- 2000 Thailand *(Giant snakes called nagas)*
- 2002 Martin Mere *(Giant catfish)*
- 2002 Cleveland *(Wallaby mutilation)*
- 2003 Bolam Lake *(BHM Reports)*
- 2003 Sumatra *(Orang Pendek)*
- 2003 Texas *(Bigfoot; giant snapping turtles)*
- 2004 Sumatra *(Orang Pendek; cigau, a sabre-toothed cat)*
- 2004 Illinois *(Black panthers; cicada swarm)*
- 2004 Texas *(Mystery blue dog)*
- 2004 Puerto Rico *(Chupacabras; carnivorous cave snails)*
- 2005 Belize *(Affiliate expedition for hairy dwarfs)*
- 2005 Mongolia *(Allghoi Khorkhoi aka Mongolian death worm)*
- 2006 Gambia *(Gambo - Gambian sea monster , Ninki Nanka and Armitage s skink*
- 2006 Llangorse Lake *(Giant pike, giant eels)*
- 2006 Windermere *(Giant eels)*
- 2007 Coniston Water *(Giant eels)*
- 2007 Guyana *(Giant anaconda, didi, water tiger)*

Other books available from
CFZ PRESS

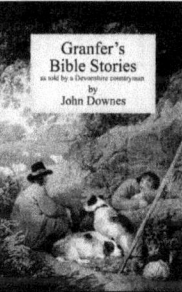

Other books available from
CFZ PRESS

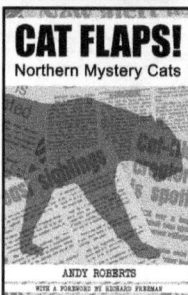

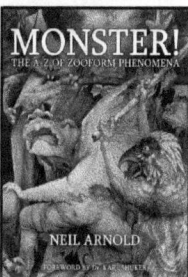

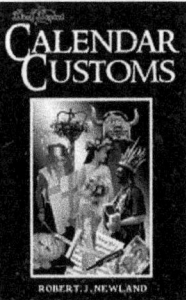